BIG CITY POLITICS

IN TRANSITION

BIG CITY POLITICS

IN TRANSITION

Edited by

H. V. Savitch
John Clayton Thomas

Volume 38, URBAN AFFAIRS ANNUAL REVIEWS

SAGE PUBLICATIONS
The International Professional Publishers
Newbury Park London New Delhi

For information address:

SAGE Publications, Inc.
2455 Teller Road
Newbury Park, California 91320

SAGE Publications Ltd.
6 Bonhill Street
London EC2A 4PU
United Kingdom

SAGE Publications India Pvt. Ltd.
M-32 Market
Greater Kailash I
New Delhi 110 048 India

Library of Congress Cataloging-in-Publication Data

Main entry under title:

Big city politics in transition / [edited by] H. V. Savitch, John
   Clayton Thomas.
         p.   cm. — (Urban affairs annual reviews ; v. 38)
   Includes bibliographical references and index.
   ISBN 0-8039-4030-0 (c). — ISBN 0-8039-4031-9 (p)
   1. Municipal government—United States—Case Studies.
   I. Savitch, H. V.   II. Thomas, John Clayton.   III. Series.
   HT108.U7   vol. 38
   [JS323]
   307.76s —dc20
   [320.8'5'0973]   91-10287
                                                                       CIP

92 93 94 15 14 13 12 11 10 9 8 7 6 5 4 3

Sage Production Editor: Judith L. Hunter

Dedication

To my wife, Marilyn, for the love she has given me and for sharing my love of big cities.

JCT

To my sons, Adam and Jonathan, whose souls were nurtured in great cities.

HVS

Contents

Acknowledgments

This book is equally the product of the two of us. It belongs, however, to the scholars of urban politics. This book was inspired by a slim volume edited by pioneering urban scholar Edward Banfield 25 years ago, and was conceived at a cocktail party among contemporary urban scholars at the Annual Meeting of the American Political Science Association just two years ago. Most of our contributors are members of the APSA's Urban Politics Section and/or the Urban Affairs Association. Not the least, the book's substance reflects decades of scholarly research and discussion by scholars affiliated with the APSA and the UAA. Two panels, one at the UAA 1991 Annual Meeting (Vancouver, B.C.), and the other at the APSA 1991 Annual Meeting (in Washington, D.C.), are enabling us to take stock of this work. We hope we have done intellectual justice to our professional associations and to our colleagues.

On a more personal note, we thank Dennis Judd and Clarence Stone for their encouragement, their good words, and their good works. At Sage Publications, Blaise Donnelly provided us with ideas for the book's readership, with editorial support, and most especially, with a panoramic view of San Francisco from atop the Hilton Hotel to inspire a brainstorming session with the book's contributors.

At the University of Louisville, Hank Savitch wishes to thank the faculties of the Ph.D. and MPA programs as well as friends in Political Science. Hank is personally grateful to Landis Jones for his support, patience, and good fellowship. Ron Vogel, who co-taught a graduate seminar with Hank, offered the right kind of skepticism at the perfect moment—just as Hank thought he was going to convince students of a dubious point. Two graduate research assistants, David Collins and Dan Sanders, did wonderful work in gathering data, editing material, and offering valuable critiques.

At the University of Missouri-Kansas City, John Thomas wishes to thank the faculty of the L. P. Cookingham Institute of Public Affairs in the Henry W. Bloch School of Business and Public Administration. Bill Eddy, Dean of the Bloch School, deserves special thanks for his support of John's effort to prepare this book while also holding administrative appointment as Director of the Cookingham Institute. Graduate assistant Dan Hoxworth also helped with library research, as well as by offering an occasional contrasting perspective on the complexities of big city politics. Finally, Lauretha Cobb, secretary to the Cookingham Institute, provided invaluable assistance—often on short notice—on many parts of the manuscript's preparation.

1

Introduction:
Big City Politics, Then and Now

JOHN CLAYTON THOMAS
H. V. SAVITCH

A QUARTER CENTURY AGO, *Big City Politics* appeared in an ambitious effort "to describe what is normal or typical about the political system" (Banfield, 1965, p. 4) of America's big cities. Author Edward Banfield sought to describe "how things really work" in each of nine large U.S. cities.

Much has changed in the years since. In the area of urban racial conflict, for example, Banfield (1965) could write:

> Although tensions everywhere are great, there has not been any outbreak of mass violence in any large city for many years. The civil rights demonstrations of the last couple of years, far from constituting exceptions to this proposition, are evidence in support of it, for although they have thrown off many sparks, none of the demonstrations has started a conflagration. (p. 13)

That generalization was obsolete by the time *Big City Politics* was published. In 1965 the Watts area of Los Angeles exploded in rioting; a year later Chicago and Cleveland fell victim to similar urban violence. Urban rioting peaked in 1967 when fires, looting, and gunshots consumed the inner cities of Detroit and Newark. Sporadic riots continued for the next few years, with even the nation's capital not spared the sight of armed troops protecting property and maintaining civil order.

Paralleling their declining ability to contain conflict, America's big cities were also losing control over their economic base. Through much of the century, cities were able to attract and hold private capital. Factories, warehouses, and the commercial fabric of the nation depended upon central cities for their vitality. When Banfield wrote *Big City Politics*, the decades of the

1940s and 1950s were the contextual references and cities were treated as if they were still the economic hub of the nation (Kantor, 1988). By the mid-1960s the economic power of America's big cities had slipped badly. Suburbs had emerged as competing poles of economic activity, drawing white-collar employment, retail trade, banking and other services. The central city was no longer the single hub of the local economy, but one of many hubs—and sometimes the weakest—in a multicentered metropolis.

Municipal fortunes were damaged further by the growing concentrations of low-income populations. Inner cities were becoming reservations for the poor, including blacks with their roots in the agrarian South and Hispanics from a number of Central and South American countries.

Until the 1960s America's big cities had functioned with relative financial autonomy, commanding a strong tax base sufficient to fund a limited range of physical services (e.g., fire, police, public works, parks, recreation). As their social-service offerings expanded in the late 1960s, cities became increasingly dependent upon state and federal aid. Eventually, as expenditures mounted to exceed revenues, many cities teetered on the edge of financial default.

The American public heard mostly about the financial plight of New York City, but New York was not alone. Well before capital markets closed on New York, the Advisory Commission on Intergovernmental Relations (1973) detailed financial emergencies facing a host of cities. Worst hit after New York were Cleveland and Detroit. Facing default, Cleveland's Mayor Dennis Kucinich lashed out at the banks, telling his constituents, "We must bring democracy to the banks as we have to our political life" (Swanstrom, 1985, p. 162). Detroit Mayor Coleman Young blamed the federal government, declaring, "New York's problems are symptomatic of a national urban crisis that has been overlooked in Washington" (Savitch, 1979, p. 179).

Writing from the perspective of the 1950s, Banfield saw urban political systems as functional, orderly, and resilient. After all, cities had absorbed wave after wave of immigration, acculturating generation after generation of newcomers and reconciling the working-class need for jobs to capital's need for reliable labor. With big cities still prosperous and stable, Banfield could say:

> Another feature that the big city political systems have in common is a remarkable ability to manage and contain the conflicts, some of them very bitter and deep-seated, that are so conspicuous a feature of the polyglot metropolis. Detroit, where workers and management, radicals and conservatives, and Negroes and anti-Negroes stand eyeball to eyeball is a good example; Atlanta is another. (Banfield, 1965, p. 13)

For Banfield's generation of scholars, this pluralism was functional for big cities, perhaps a bit sloppy but never an obstacle to problem resolution (Sayre & Kaufman, 1960; Banfield & Wilson, 1961; Dahl, 1961). No one worried that this conflict might run rampant and weaken city government. A new generation of scholars emerged to challenge this outlook in the 1970s. Reflecting on the new shape of city politics, Yates (1977) contended that America's big cities had become ungovernable. In his view, a tolerable pluralism had degenerated into an unmanageable "street-fighting pluralism," that is, "a political free-for-all, a pattern of unstructured, multilateral conflict in which many different combatants fight continuously with one another in a very great number of permutations and combinations" (Yates, 1977, p. 34).

Some contemporary scholars agree. Far from being manageable, America's big cities may be struggling, intractable political entities whose governing mechanisms are crumbling. Epidemics of crime and drugs now engulf many cities, their educational and criminal justice systems are failing, and the social structure of the poor is disintegrating. In at least some cities, problems may have outrun the ability of local governments to cope.

Is the picture so bleak everywhere? Obviously not. Some cities continue to manage conflict and induce cooperation. As one well-known example (Stone, 1989), Atlanta is still held together by interacting government and business elites. In a somewhat different manner, Seattle still functions relatively effectively. As another positive sign, some big cities are no longer willing to surrender their terrain to indiscriminate development. Such cities as San Francisco and Boston have combined governmental reform with changes designed to protect and enhance the environment.

America's big cities obviously have changed greatly, evolving in ways no one could foresee, in the almost 30 years since Banfield wrote. The time may be right to put those changes into context. Drawing our inspiration from Banfield, we use this volume to describe, analyze, and explain how big city politics has been transformed in the last three decades. Like Banfield, we examine "how things really work" in the contemporary big city.

Unlike Banfield, we have the considerable advantage of his earlier work as a base point in comparing how cities have changed. In addition, the hindsight Banfield provides permits us to pay greater attention to patterns and variations between cities.

These comparisons are facilitated by the rich and able cast of scholars who have contributed to this volume. Those scholars provide profiles of eight cities contained in Banfield's study—Atlanta, Boston, Detroit, Los Angeles, Miami, Philadelphia, St. Louis, and Seattle. Other scholars contribute profiles of five cities—Chicago, Denver, Houston, New Orleans, and San Francisco—added to enhance the regional balance in this national portrait.

With these additions, the cities may be compared not only by Sunbelt versus Frostbelt, but also as Eastern (Boston and Philadelphia), Midwestern (Chicago, Detroit, and St. Louis), Southern (Atlanta, Miami, and New Orleans), Western (Denver and Houston), and Pacific Rim (Los Angeles, San Francisco, and Seattle).

Each city profile addresses topics similar to those in the original *Big City Politics* but with allowances for the great changes of the last three decades. Our inquiry highlights (a) the demographics and economic base of the city; (b) the role and structure of city government, including interaction with state houses, suburbs, and Washington, DC; and (c) the cast of interest groups and political influentials. Cutting across these concerns is a general interest in the political character of the city—the composition and cohesion of the coalitions, groups, organizations, and individual actors that shape major decisions. This assortment of forces may participate in either formal or informal arenas of decision making and may include chambers of commerce and business associations, labor unions, neighborhood groups, local politicians, and officials from other governments (e.g., governors, state and national legislators). In what follows in this chapter, we set parameters for the city profiles by describing some of the principal changes in big city politics since Banfield wrote.

DEMOGRAPHICS AND ECONOMICS

America's big cities are notable first for the number of people who left them over the last 30 years. Although, as Table 1.1 shows, our 13 cities as a group lost only 5.5% of their population between 1960 and 1990, dropping from slightly more than 15 million people to around 14 1/2 million, a number of cities were hit very badly. St. Louis lost nearly half of its population, Detroit lost more than a quarter, and Philadelphia and New Orleans both fell by 21%. Population growth was limited to a few Sunbelt and Pacific Rim cities, topped by Houston which, with the aid of annexation, increased by nearly 75%.

More significant than the sheer numbers were the substantive changes in population. The 1960s marked the beginning of a new era in urban diversity. Although Banfield and his contemporaries were well aware of the migration of Southern blacks to Northern cities, they did not foresee the coming of substantial numbers of Hispanic and Asian and Pacific Island settlers. The Hispanic migration—from Mexico, Puerto Rico, Central America, and parts of South America—reached unprecedented levels in the 1970s and 1980s. By the 1980s some demographers were predicting that Hispanics eventually would succeed blacks as the largest minority group in the United States.

TABLE 1.1
Population Shifts in Thirteen Central Cities: 1960-1990 (in thousands)

City	1960	1990	% increase/decline
Atlanta	487	394	-19.1
Boston	697	574	-17.7
Chicago	3,550	2,784	-21.6
Denver	494	468	-5.3
Detroit	1,670	1,028	-38.4
Houston	938	1,631	73.9
Los Angeles	2,479	3,485	40.6
Miami	292	359	22.9
New Orleans	628	497	-20.9
Philadelphia	2,003	1,586	-20.8
St. Louis	750	397	-47.1
San Francisco	740	724	-2.2
Seattle	557	516	-7.4
Total	15,285	14,443	-5.5

SOURCE: U. S. Bureau of the Census. (1962). *1962 statistical abstract of the United States,* Table 14, pp. 22-23. Washington, DC: U. S. Government Printing Office. U. S. Bureau of the Census. (1990). *1990 statistical abstract of the United States,* Table 40, pp. 34-36. Washington, DC: U. S. Government Printing Office.

As always, big cities were the first to experience the cultural and political impact of immigration. Miami (Cubans), Los Angeles (Hispanics, Asians), San Francisco (Asians, Pacific Islanders), and New Orleans (Cubans, Central Americans) now resonate with a new ethnic tone. The tone also can be heard in Chicago, Denver, Houston (all Hispanic) and, in more muted form, in most other large cities.

The several migrations resulted in minorities—blacks, Hispanics, and/or Asians—achieving majority population status in many big cities. In some (Atlanta, Detroit, and New Orleans), blacks became the majority; in others (Miami), Hispanics were the majority; still others (Los Angeles) became so-called "majority minority" cities, with a combination of minority groups holding majority status. In almost all large cities, newly arrived immigrants were a force to be considered. Table 1.2 suggests the magnitude of that force even by 1980.

The new migrants arrived even as most big cities were seeing an industrial evacuation. During the late 1960s and through the 1970s, in the Northeast and Midwest in particular, commerce and jobs followed the white middle class to the suburbs. Some economic hope could be found in the efforts of City Hall to attract services, tourism, and office employment into downtown centers. Local elites attracted developers to construct new skylines and to

TABLE 1.2

Minority Populations in Thirteen Big Cities: 1960-1990 (as % of total population)

City	Percent Black		Percent Hispanic	
	1960	1990	1970	1990
Atlanta	38.3	67.1	1.0	1.9
Boston	9.1	25.6	2.8	10.8
Chicago	22.9	39.1	7.4	19.6
Denver	6.1	12.8	16.8	23.0
Detroit	28.9	75.7	1.8	2.8
Houston	22.9	28.1	12.1	27.6
Los Angeles	13.5	14.0	18.4	39.9
Miami	22.4	27.4	45.3	62.5
New Orleans	37.2	61.9	4.4	3.5
Philadelphia	26.4	39.9	1.4	5.6
St. Louis	28.6	47.5	1.0	1.3
San Francisco	10.0	10.9	14.2	13.9
Seattle	4.8	10.1	2.0	3.6

SOURCE: U. S. Bureau of the Census. (1977). *County and city data book*, Table 4. Washington, DC. U. S. Government Printing Office. U. S. Bureau of the Census. (1979). *County and city data book*, Table C. Washington, DC: U. S. Government Printing Office. U.S. Bureau of the Census, telephone interview, March 28, 1991.

remake old waterfronts. But these efforts fell short of matching in quantity, quality, or salary earlier jobs in manufacturing and skilled trades.

For a time, some cities in the Sunbelt and on the Pacific Rim successfully resisted the gravitational pull of urban decline. During the 1970s such cities as Atlanta, Houston, and Los Angeles enjoyed an economic boom, in Houston's case helped by liberal annexation policies that permitted the incorporation of suburban growth.

Meanwhile, many of the declining cities found temporary relief in federal assistance. Beginning with Lyndon Johnson's Great Society, federal urban programs mushroomed between 1960 and 1969 from a few dozen doling out several billion dollars, to more than 500 providing over $14 billion (Goldfield & Brownell, 1990). Richard Nixon's New Federalism sought to replace the Great Society with fewer strings and greater discretion, offering cities general revenue sharing, special revenue sharing, and block grants. Jimmy Carter sought to better target existing programs and to develop competitive grants-manship in cities, offering training programs for the unemployed and Urban Development Action Grants (UDAGs) to stimulate local economies.

All of this help may have softened the blows of urban decline but did little to reverse them. Cities began to treat federal funds as just another general revenue source rather than as a means to create new industry and to resolve chronic problems. In effect, federal funds were used to fill temporarily the economic holes left by departing industry, a choice that exacerbated the urban fiscal crisis of the 1970s.

The risk in that choice became evident when federal funding began to decrease, gradually in the late 1970s and then dramatically under Ronald Reagan. By the mid-1980s Reagan had delivered the *coup de grace* to federal aid to cities. As supply side economics took hold, the White House cut funding for housing, community development, mass transportation, and public works. The Reagan approach to urban decline relied upon private rather than public investment, and the economic watchwords were *privatization* (of public services) and *tax incentives* (for development).

With federal funds in short supply, America's big cities in the 1980s began to compete vigorously with each other for economic growth. The new competition frequently found cities bidding against each other with all manner of economic inducements to lure industries to their towns. On the one hand, the revival of a city such as Boston suggested that the right strategy might produce success. On the other hand, the abrupt fall of much-touted Houston suggested that economic fortunes might reflect only cyclical fluctuations in the international economy.

In the face of such turnarounds, the once-distinct Sunbelt-Frostbelt regional contrast blurred. Scholars, journalists, and policymakers had argued once that American urban dwellers should follow the sun, leaving behind the outmoded, tired, and declining Frostbelt cities of the Northeast and Midwest, to escape to the seemingly clean, efficient, and prosperous Sunbelt cities of the South and West. Time and closer examination revealed that contrast to be simplistic. The urban crisis was not just regional; cities in the Frostbelt could revive and prosper, and those in the Sunbelt might fall on hard times.

ELECTORAL AND GOVERNMENTAL STRUCTURES

In a time of many dramatic changes, the governmental structures of America's big cities remained largely intact. The charters under which most big cities operate were not dramatically altered. Today's urban officials attain and manage their offices in much the same manner as their predecessors did 30 years ago.

Of the structural changes that were made, the most significant affected electoral rules. Federal voting-rights legislation spurred this change by calling into question any electoral rules that adversely affected the voting power of minorities. Aided by that legislation in their desire to represent better the new ethnic diversity, reformers pushed many big cities to elect more city council members from districts rather than at large. Traditionally favored by reformers as the best way to reflect the public interest, at-large elections worked to the disadvantage of minorities who tended to be concentrated in distinct neighborhoods. Their majority status in those neighborhoods gave blacks, Hispanics, and Asians a good chance of winning district elections but little chance of winning on a citywide at-large basis where, in the 1960s and 1970s anyway, they were still a minority.

The push for electoral reform, however, affected only a fraction of America's big cities. Most cities already elected at least some council members from districts, and a few, such as Detroit, were reluctant to abandon their at-large elections.

The structure of city government was even more resistant to change. Ethnic diversity tested the council-manager system, which operates best in homogeneous, consensual environments. But few big cities that used the system in the 1960s have abandoned it. One expert recently estimated that, of all 3,000 city manager municipalities in the United States, only about 10 consider abandoning the system in any given year, and only 1 or 2 actually make the change (Mobley, 1988).

There was no rush by large mayor-council cities to adopt the council-manager plan. None of the cities in this volume chose to shift the center of political gravity from their mayors to appointed managers.

Despite the resistance to altering the legal forms of government, the practical demands of running cities required sweeping changes in habit and incremental changes in administration. Across America's large cities, there was a clear convergence between what reformers had long preached as professionally correct and what traditionalists had long recognized as political reality. From the one side, many unreformed cities moved toward reform principles by turning to appointive chief administrative officers (CAOs) to head their bureaucracies. With the help of the courts, municipal patronage continued to decline. From the other side, many reformed cities saw their city managers lose power to politicians, mayors in particular. As part of that evolution, most large council-manager cities now elect mayors directly, rather than having the council make that choice from among its membership as traditional reformers favored.

Mayoral power and visibility have grown, with many big cities run by high-profile mayors in the years since Banfield wrote. The range and depth of this change are substantial—John Lindsay and Ed Koch in New York, Richard Daley, Jr., and Harold Washington in Chicago, Dianne Feinstein and Art Agnos in San Francisco. Minorities are prominent among this new breed of mayors, with representatives such as Detroit's Coleman Young, Los Angeles's Tom Bradley, Atlanta's Maynard Jackson and Andrew Young, Denver's Federico Peña, and Seattle's Norm Rice.

New mayoral power derives from several factors. The increasing importance of the mass media, in tandem with the declining role of political parties, favors personalities and individuals over issues and teams. Increased ethnicity also demands centralized political leadership to manage and resolve conflict. Highly charged political climates are hostile to assertions of managerial authority but conducive to political leadership. The pivotal role of mayors in securing federal help and in luring developers to their cities also has strengthened the office.

The 1960s anticipation that a new tier of metropolitan government would rise over closely situated municipalities has been disappointed. Megapolis has assumed its demographic shape without spawning a political cover. Today the call for metropolitan governance is faint, notwithstanding the accomplishments of metropolitan governments in Minneapolis-St. Paul, Indianapolis, and, to a lesser extent, Miami. Beneath the surface, however, cities have made incremental accommodations through city-county compacts, through regional planning or coastal commissions, through different types of public benefit corporations (mostly in development and transportation), and by relying upon state governments for coordination.

Is structural reform no longer a major force in big city politics? The old debates about reform certainly have faded, if not disappeared. Big city governments no longer struggle with whether to professionalize the civil service. The resilient impulse toward reform in big city politics still could survive, however, if perhaps with new targets appropriate to a new era.

INTEREST GROUPS AND POLITICAL INFLUENTIALS

When Banfield wrote, America's big cities were notable for the relative quiescence of political groups. In contrast to eras before and after, the period of the late 1950s through the early 1960s found most big cities with relatively few active or influential political groups.

Big city politics in this era was dominated by organized business interests. In unreformed cities, political parties collaborated with business elites, often with the support of neighborhood or union interests, in running cities. In some unreformed cities, the collaboration was directed by a political machine that served as an integrative mechanism through a process of discrete and divisible accommodation—always making sure that private investment could be absorbed by public works and by the growth of public jobs.

In reformed cities, business often held an even more powerful position, working directly with candidates and sometimes recruiting them from within the business community. Other principal political groups were reform, "good government" groups that were themselves strongly business influenced. Discouraged by the citywide emphasis of the reform structures, other political groups were notable more for their absence than their activity.

This relative political calm was broken by the urban riots of the 1960s, which signaled the beginning of an expansion in the number, power, and stridency of political groups in America's big cities. Greater diversity provided the principal impetus for this expansion as growing urban ethnic populations, first blacks and then Hispanics, formed new political groups to represent their interests. In Kansas City, Missouri, for example, a group called Freedom, Incorporated, formed in 1962 to represent the political interests of the city's growing black population and then quickly became one of the most influential groups in both the city and the surrounding county (Fitzpatrick, 1987).

The growing minority presence also indirectly stimulated the growth of new groups in predominantly white areas of cities. Feeling threatened by racial turnover, whites who had resisted the move to suburbia formed neighborhood organizations to preserve the stability of residential enclaves. In Cincinnati more than a dozen new neighborhood organizations appeared in the late 1950s and early 1960s in areas threatened by racial change (Thomas, 1986). Similar waves of new neighborhood organizations appeared in Philadelphia, Boston, Chicago, and, in varying degrees, throughout the country.

The federal programs of the 1960s and 1970s nurtured the growth of both minority and neighborhood organizations. Black political groups drew sustenance from the Community Action and Model Cities agencies established during the Johnson era. In the 1970s the Community Development Block Grant program gave a similar boost to the clout of neighborhood organizations.

As the power of insurgent groups grew, the clout of some traditional groups declined. Business interests, for example, at least temporarily lost some of their power at the peak of federal influence in the early 1970s. In what may be a more lasting change, political parties steadily weakened, perhaps in part because they failed to absorb and represent the new group interests.

HOW IT REALLY WORKS: HYPERPLURALISM OR URBAN REGIMES?

The many changes of the last 30 years inevitably affected how politics works in America's big cities. In elections, with the weakening of political parties, the new ethnicity has proven a powerful organizing tool. Harold Washington in Chicago demonstrated how ethnic neighborhoods could be mobilized for a candidate. In Philadelphia and New York, contests have been fought largely between ethnic/racial blocks.

In policy debates, the thrusts and counterthrusts of the new organizations have sharpened the dichotomy between City Hall and neighborhoods, between downtown and popular interests, between progrowth and antigrowth coalitions. New fault lines have developed, with interest groups clustering on either side. Conflicts have erupted over developmental issues (airports, highways, high-rise buildings), over the delivery of city services, over the placement of incinerators, and over issues of crime and drugs. The slogan that epitomizes the struggle is "Not in My Backyard!" or NIMBY (Open a methadone clinic? Not in my backyard!). NIMBY articulates a collective protest against the disruptions of urban life and related intrusions of city officials.

The processes of governance have also changed, but in ways not as well understood. Clearly, running a city today, to a much greater extent than 30 years ago, requires assembling a coalition of political groups. What is not clear is how often contemporary big cities succeed in producing these coalitions. It also is not clear, given the uncertainty about that success, whether the segments of the population that benefit most from the decisions of urban governments are the same today as 30 years ago.

In answer to the first question, Yates (1977) argued more than a decade ago that success in building effective governing coalitions may be relatively rare in the "ungovernable city." In his perception, nobody really makes decisions; or, if decisions are made, they are too convoluted to discern a pattern of winners and losers. With more interest groups and more issues, contemporary big city politics may be most notable for its incoherence and ineffectiveness:

> Policy outcomes are a product of highly variable fragmented and unstable problem and policy contexts. And it is precisely because urban policy makers must deal with so many different, fragmented problem and policy contexts that urban policy making as a whole is so fragmented, unstable, and reactive. (Yates, 1977, p. 85)

This *hyperpluralistic* tendency may be magnified by the declining importance of the central city within the larger metropolis. With cities losing population and economic vitality to the scattered suburbs, the locus of political power may be disappearing. The result could be a centrifugal, undefinable, and ungovernable metropolis where decisions splay in many directions and add up to little.

A contrasting vision of the city pictures amalgams of coalitions working through informal processes to produce coherent decisions. In the *urban regime* model, as Stone (1989) documented in Atlanta, the tendency toward fragmented political power can be resisted successfully by an effective governing regime. That regime is essentially a ruling majority coalition composed of several political fragments, including a dominant business leadership group and elected political leaders aligned with enough other groups to produce an electoral majority. The coalition is maintained by providing particularistic benefits or "side payments" to each of the partners in the coalition.

Urban regimes in the 1990s may differ in some respects from their predecessors of the 1960s. In particular, the new ethnicity may have forced business groups to turn to new coalition partners. In decades past, business groups often obtained their electoral majorities by joining with the white middle class. Today, it is not uncommon for business to form alliances with black constituencies, as occurred in Atlanta and Detroit. In other cities the coalitional partners might include neighborhood organizations, labor unions, or other groups.

Some cities may fall somewhere between the extremes of hyperpluralism and an effective governing regime. That middle ground might be occupied by an ad hoc majority coalition (e.g., electoral coalition) that is more organized than hyperpluralism implies but lacks the longevity necessary to constitute a regime.

The beneficiaries of big city politics may vary, depending upon what type of governing regime or hyperpluralism typifies America's cities today. Business leaders probably will benefit in either case, although perhaps more in a structured governing regime than in the less predictable laissez faire of the ungovernable city. Other groups may partake of benefits, depending upon their ability to establish viable alliances with business interests.

The models differ clearly on the fate of the central city. For the hyperpluralist, fragmentation and multisided conflict work against rational policy-making and ultimately contribute to the further disintegration of the city. For the regime theorist, the city may be capable of exacting benefits from particular coalitions, but less organized and less powerful constituencies could be given short shrift.

In the following pages we will see how these models hold up. Each profile examines a particular city with an eye toward what kinds of groups, coalitions, or actors have emerged since the 1960s and what that portends for big city politics in the 1990s.

REFERENCES

Advisory Commission on Intergovernmental Relations. (1973). *City financial emergencies.* Washington, DC: Government Printing Office.

Banfield, E. C. (1965). *Big city politics.* New York: Random House.

Banfield, E. C., & Wilson, J. Q. (1961). *City Politics.* New York: Vintage.

Dahl, R. (1961). *Who governs? Democracy and power in an American city.* New Haven and London: Yale University Press.

Fitzpatrick, J. C. (1987, September 15). Freedom: A force to reckon with. *Kansas City Times,* p. A1

Goldfield, D. R., & Brownell, B. A. (1990). *Urban America: A history.* Boston: Houghton Mifflin.

Kantor, P. (1988). *The dependent city.* Boston: Little, Brown.

Mobley, J. (1988, February). Politician or professional? The debate over who should run our cities continues. *Governing 2,* 42-48.

Savitch, H. V. (1979). *Urban policy and the exterior city: Federal, state and corporate impacts upon major cities.* Elmsford, NY: Pergamon.

Sayre, W. S., & Kaufman, H. (1960). *Governing New York City.* New York: Russell Sage.

Stone, C. N. (1989). *Regime politics: Governing Atlanta, 1946-1988.* Lawrence, KS: University Press of Kansas.

Swanstrom, T. (1985). *The crisis of growth politics.* Philadelphia: Temple University Press.

Thomas, J. C. (1986). *Between citizen and city: Neighborhood organizations and urban politics in Cincinnati.* Lawrence, KS: University Press of Kansas.

Yates, D. (1977). *The ungovernable city: The politics of urban problems and policy making.* Cambridge, MA: MIT Press.

2

Boston:
The Incomplete Transformation

PHILLIP L. CLAY

THE LAST 30 YEARS have been a period of great change for Boston. The periods before and after the 1960s present sharply different images (Banfield, 1965; Ganz, 1987; Doolittle, Jackson, Masnick & Clay, 1982; Thernstrom, 1969; Clay, 1989a). The Boston that Banfield wrote about in the mid-1960s faced population decline; the new Boston has some recent population growth. Old Boston saw a decline in its middle-income population; new Boston has experienced a growth in its middle-income and upper-income populations. Boston's former in-migrants were mainly blacks from the South; new Bostonians are newly arrived Southeast Asians, Irish, and South and Central Americans. Old Boston wondered if any jobs would be safe, given the great economic hemorrhaging that resulted from the decline of traditional manufacturing; new Boston has had real growth on a diversified economic base.

Old Boston was timid and took small steps or major gambles in reviving its downtown; new Boston is held up as a model of both the process of dynamic downtown development and strategic economic revitalization to achieve a diverse base. Old Boston found the Irish in firm control and others in an uneasy accommodation to them; new Boston is led by a broad, progressive base of leadership not just in City Hall but increasingly in its civic and private spheres as well. Old Boston's economic elite (i.e., lenders, corporate and insurance executives, and professionals such as lawyers) disdained dealing with city problems; the new elite (composed of some of the same types but including entrepreneurs, academics, and other new players) is active in broad partnerships to deal with housing, education, and jobs.

The change in Boston is not the result of a discrete or dramatic event or of revolutionary change. It is not the result of a charismatic personality or a singular economic event. While important individuals and fortuitous events played a part in this transformation, the change is at its roots the result of the playing out of basic urban change processes and the rebalancing of political and economic interests (Clay, 1989a): the reader should view these trends as interactive, not unidirectional.

A NEW ECONOMIC ROLE AND A NEW URBAN FACE

Boston, like other major cities, lost its major economic role nationally and within the region when traditional manufacturing (e.g., leather goods, textiles, machines) took its great slide between 1920 and 1950. Because Boston's business and professional services catered to this sector, they experienced decline as well.

What changed for Boston and for other cities in the 1970s was its early and effective experience in riding the wave of downtown business, retail, and professional services associated with high-tech development, medical research and development, and computer services. Boston and other cities took advantage of these economic development activities in a major way and reinforced these with a substantial expansion of its banking and financial service activities to become one of the nation's major money centers (Frieden, 1989). This expansion buoyed a construction and development sector. Boston also became a major center for export of advanced business services (e.g., consulting, advertising, computer software, specialized legal practice). The result was a substantial expansion of its economic base and the kind of diversification that allowed the city's economic expansion to continue through the 1980s (Ganz, 1987; Boston Redevelopment Authority, 1989; Yudis, 1984).

Boston was successful in creating jobs at both ends of the income spectrum. On the one hand, there was a substantial expansion of professional, administrative, and managerial positions attracting young workers from the suburbs and other parts of the country. White collar job growth in this area was roughly a third of the population growth (50,000 net) the city experienced in the early 1980s. On the other hand, while growth in this area was significant, it was not nearly as large as the growth in the lower status, lower paying service jobs, which attracted immigrants to fill positions that nearly fully employed Bostonians found less attractive. These workers would staff fast-food outlets and clean the hotels and office towers that became the physical representation of the new Boston.

Development in Boston was complementary to the more substantial expansion of jobs and enterprises in suburban areas and in the hinterland. Growth outside the city would require the services of venture capitalists, patent attorneys, and software engineers who preferred to concentrate in the city. While these rapid-growth areas outside the city became major sub-centers, producing agglomerations that in other parts of the country would be considered small cities, both the core and the periphery were supported by state policy that was clearly prourban and aggressive in promoting diversified economic development (Ferguson, 1986).

Over a two-and-a-half-decade period, Boston developed two faces. First, it became a high-wage economic dynamo in a region that bucked the secular regional trend toward decline. Wealth and economic clout in some of its people increased as evidenced, for example, by its soaring housing prices and the expansion of specialty retail and personal services in the city.

Second, Boston became a city where a growing number of service workers and discouraged young people with persistent poverty defines the "urban underclass" (Osterman 1989). With this contrasting wealth and poverty would go opportunity and isolation, public promotion of development and demands for public relief from such negative impacts as homelessness.

The physical changes promoted by economic development and population growth triggered several types of competition. The transformation of Boston, unlike that of cities in the South and West, did not take place in the context of vast expanses of open land, the annexation of prosperous areas at the city's margin, or by new players developing land that had been overlooked. Boston was fully developed and surrounded by fully incorporated areas. While there were areas of decay, abandonment was limited to a few areas.

The physical transformation was framed by competition for the right to use or reuse a variety of land areas and related resources, including regulatory power. Each new use—whether an office building, an institutional expansion, or a housing complex—had to displace an existing use. Each case of competition for the right to use partly reflected the relative influence of various private interests, public goals (sometimes), and the balance between them. Over the last three decades several of these arenas of competition have developed.

The first major competition was for space in the central core, including the struggle over who benefits from the expanding development opportunities, what visions of downtown prevail, and who decides design. Another struggle was over the issue of to what degree preservation of specific buildings and urban scale was a priority. The winner was a strong coalition for downtown growth (Mollenkopf, 1983). While there were exceptions, developers often put up buildings that were poorly massed, darkening to the streets below, and devoid of features that connected to the fabric of the district. Preservationists lost, as did those who preferred to keep the traditional sense of scale in the downtown. They would be overruled largely by political and economic forces until the mid-1980s.

The second major competition was between the forces who wanted new development (at the price of the loss of old neighborhoods) and those advocating preservation of existing neighborhood scale and urban scale (even at the risk of putting a damper on private development initiatives). There were winners and losers in this struggle as well. The extremes are illustrated by what happened in the West End in the early 1960s, compared with what happened in the South End 10 years later (Keyes, 1969). In the former case, an old but viable neighborhood was demolished to make way for a set of sterile middle-income rental towers and the expansion of a major hospital (Gans, 1965). The loss of the West End and the social dislocation it caused led to the vigorous defense and subsequent revitalization of the South End neighborhood (Auger, 1979).

Other areas—Charlestown and Roxbury, for example—would experience a more gentle urban renewal (Keyes, 1969). This seemingly modest drama would ignite political passions for a generation, and the lessons would be the progressive core in 1980s politics (King, 1981; Lupo, 1981).

In areas that saw modest physical change and development, the competition was between incumbent residents and newcomers over the nature of the changes that would occur, who would decide, and who would benefit. This distinction was framed as competition between the poor and the nonpoor. The struggle produced a substantial number of winners and losers. For example, in the late 1970s and early 1980s, speculators won, but their substantial victory then under a prodevelopment mayor would set them up as targets for the new regime in the 1980s.

In gentrified areas, especially those close to the city center (and those on the "hills" of Boston where nineteenth-century Victorian homes are prominent), the poor were in direct competition with the middle class for housing by the mid-1970s. Housing that formerly filtered down to the poor as city dwellers fled the city was now filtering up to middle-income newcomers who faced slow building in the suburbs or who found Victorian architecture increasingly attractive.

In that competition, the poor would lose and would lose again in the late 1980s as one seventh of the nonsubsidized rental housing stock was converted to condominiums. The loss in rental units translated into higher rents and prices and into a very low vacancy rate (Clay, 1989b). To a greater degree than in just about any other city, except perhaps San Francisco and New York, housing would be viewed as a major area for political reaction and policy action. These forms of competition were not entirely new. Some were perennial issues, others were special to this era. What is significant is that they occurred together in Boston and helped define how a growing economic pie was sliced. As the pie increased, the inequalities would either be more visible or less tolerable or both. The interests of winners and losers would become clearer in political terms.

ADAPTING TO DEMOGRAPHIC CHANGE

In 1950 the population of Boston was 801,000. The population reached a low of 562,000 in 1980. Between 1980 and 1985, the population grew to an estimated 601,000. This 7% increase in population in five years included a broad range of newcomers and was net of the continued out-migration of working-class and lower middle-income whites. The newcomers were largely immigrants and white-collar workers who would invigorate declining neighborhoods and would consume the units that were left by outmovers, converted to condos, or rehabilitated (Clay, 1989b). This growth in the population changed the demand for services in neighborhoods and in the core, strengthened the tax base, and supported a host of private enterprises and services that had not been viable before.

Another major change occurred in the size and composition of the poor population. The poverty population decreased marginally from just over 20% to just under 20% (Osterman, 1989). More important than the modest change in size of the poverty population are the major compositional shifts. The very poor, female-headed families and the persistent poor form a growing share of the low-income population that formerly were dominated by elderly and single persons, and nuclear families who were temporarily in need. Now nuclear families are a small minority of the poor and appear to have benefited by economic development. This new poor population is not helped and is unlikely to profit from economic shifts or ordinary efforts at job training or community economic development.

With the changing composition of the population has come an increase in the number of people of color. A decline in the white population means that the share of the population represented by blacks, Hispanics, and Asians went from less than 10% in 1960 to more than 36% in 1990. While blacks represent the largest nonwhite group, the Hispanic and Asian populations have been the fastest growing in recent years (Boston Redevelopment Authority, 1989; Kaufman, 1984).

This population growth and shift in class groups (a larger share of the poor are minority) have made school and residential segregation issues sharper and have highlighted the continued marginality of people of color in the city. The schools are nearly totally segregated and of poor quality. While not as sharp as in the 1960s, residential segregation is still an important factor (Clay et al., 1985).

Fewer than 20% of the total households in the city have children in the public schools. Such an expanding minority population that dominates enrollment means that, increasingly, public education is a matter of race. This issue has been noted in the selection of a school superintendent and in school budget reviews.

The discovery of education as a key element in economic development has made educational quality a critical issue. The demographic trends related to race have stimulated a new round of civic organizations and business interests operating parallel to the political structure. This stimulation is based on a growing recognition that the labor force will be increasingly minority. Good education is linked to regional economic competitiveness.

In the midst of these trends, the middle class has grown substantially; high-income groups expanded from 26% to 36% of the population (Kluver, 1988). These groups paid more for housing, and within a short period of time rents more than doubled. Less well-off households were required to bid higher and in some cases were displaced (Auger, 1979). Middle-class interest fueled the demand for the private apartments that were converted to condo-miniums. The presence of middle- and upper-income households fundamen-tally changed neighborhoods and altered the social and commercial fabric. Neighborhood bars became chic, and Laundromats were replaced by wine-and-cheese shops. There would be less tolerance for crime and more demands for such public amenities as tree planting, jogging paths, and gas streetlights.

The low-income population displaced from areas they could afford became even more concentrated in enclaves whose residents were mired in persistent poverty (Clay et al., 1985). The middle class (mainly singles and childless couples) became an important interest group. They would vote, would be at least moderately progressive, and would support initiatives to empower neighborhoods. Middle-class presence would be a powerful source for political change so long as class interests were muted. Because the 1980s were prosperous, there were fewer hard choices and more muted conflicts of interest. A slowing economy will present opportunities for making hard choices.

Boston's transformation also comprises a rise in small or single-person households. The small household is a permanent phenomenon that may expand further and be the basis for continuing change. This type of household ought to be viewed against the city's racial patterns, and the trend portends a city split between mainly adult white neighborhoods and large-family minority neighborhoods.

POLITICS AND THE TRANSFORMATION OF BOSTON

Prior to the 1920s, Boston and the Commonwealth of Massachusetts were run mainly by Americans of English ancestry, often called "Yankees" (Whitehill, 1968). Slowly, as the Irish gained in demographic dominance and power, they elected a mayor. Mayor James Michael Curley established the first Irish political machine that ran the city between 1917 and 1949 (Levin, 1960). During his tenure, the battle lines were drawn: the Irish (Democrats) were in firm control of the city, and the Yankees (mainly Republicans) controlled state government. Both sides reinforced their own power.

The Irish created a structure in the city (i.e., nonpartisan primaries, strong mayor, at-large elections) that assured other groups (Italian-Americans, Yankees, Jews, blacks, etc.) would have little chance to unseat them (Boston Urban Study Group, 1984).

The Yankees reinforced their power. They controlled the economic sphere by the control and ownership of the wealth, financial institutions, and large companies. They also controlled major law firms, the courts, the newspapers, and organs of informal influence. Their interest in city government was simple: they did not want municipal authority to get in their way or themselves to become victims of the corruption for which the city was becoming well known.

The Yankees also exercised increased levels of control over city government. Exercising their power through the state, they controlled taxation and an array of powers normally left to cities, such as appointment of a police chief and control of the Port of Boston, local courts, water and sewer departments, auditing and financial oversight, mass transit, and civil service. These and other powers were placed into the hands of the state agencies directly or into the hands of public authorities over which the legislature and/or governor had direct control.

Between 1920 and 1965 the struggle for Boston was viewed as a game of who controls and arranges the fading and tattered deck chairs on an urban *Titanic*. Boston experienced a steep economic decline. The city lost a quarter of its population by 1970; not one new major building was constructed in the city between 1920 and 1960; and jobs and population moved to the suburbs. City planning and economic development were largely *pro forma* exercises (Boston Redevelopment Authority, 1965). Yankees invested their money outside of Boston, and the Irish dug in to protect the control they had won. This unwritten contract on spheres of control defined influence. The completion of a circumferential highway around Boston in the mid-1950s seemed to seal the city's fate as the "suburban noose" became broader and stronger.

The city ran on this political contract until the 1960s. In 1959, Boston elected the first of a series of administrations put together by a coalition for growth and transformation that would last until 1983 (Mollenkopf, 1983). This was a new covenant to replace the old standoff between the Irish and the Yankees.

While the Irish still dominated city politics during the 1960s and 1970s, this period marked the ascendance of more politically broad-minded and ambitious Irish mayors (sometimes referred to as "lace-curtain Irish") who actively sought to build bridges to the business community while doing what was necessary to hold on to their narrowly ethnic constituency.

This political coalition was replaced in the election of 1983 with a new contract that reflected the view that the old understanding should be rewritten to include new groups such as blacks and the new middle class. A city administration with a conservative Democratic tradition now developed policies that were viewed as radical (e.g., rent control, exactions from big developers to contribute to neighborhood projects, gay rights). In rewriting, special attention was given to who participates and what the substantive obligations of city government were. We will call the former phase (1960 to 1983) the transformative period and the second phase (1983 to the present) the progressive period. Each of these eras is discussed below, but first a note about race.

As early as the mid-1960s, race was a key political issue in Boston. Each administration had to address the ascending demographic and political power that blacks and then Hispanics would exhibit. A critical mass was reached by the late 1970s, although blacks had delivered key blocs of votes in elections before this date.

The tensions reflected over the last 30 years and in the transition to a new regime in the 1980s produced anomalies and interesting twists in Boston politics. The same voters who gave the nation John Kennedy and Edward Brooke also cheered Louise Day Hicks, the heroine of antibusing activities. A state representative and significant player in the antibusing drama in the mid-1970s, Ray Flynn became the populist elected mayor a decade later on the promise to heal the racial wounds in the city and to make blacks a partner in a city that had shown little respect to them. Even with various dramatic events, black efforts to be counted and included met with frustration (Lupo, 1981; Lukas, 1985). Racial aspects and other issue are discussed for each era below.

THE TRANSFORMATIVE PERIOD, 1960-1983

In the late 1950s, it was agreed widely that Boston's effort at urban renewal in the postwar period had been a failure. While a depressed business community had few expectations about what the city could do, it was tired of the corruption and political wrestling that came with doing business in the city. It also heard about the urban-renewal projects being mounted in other cities. Faced with a mayoral election in 1960 and the prospect of a strong prolabor and ethnic candidate winning that contest, business embraced John Collins, an Irish lawyer. Without an organization or a political history, Collins offered a departure from "politics as usual." In a political upset, Collins was elected by a narrow margin. He put together a program of development in the city that brought unlikely characters into an active relationship. Bankers, developers, state officials, liberal reformers, and academics joined Collins and his development director, Ed Logue, in putting together and implementing one of the most ambitious urban-renewal programs in the nation. Early projects emphasized downtown development and expansion of institutions. Other projects would renew neighborhoods, first in the highly destructive manner common in such projects but later in ways that would take more notice (if not full account) of residents' feelings and interests. Projects over time gave more attention to renewal rather than removal. The "planning with people" that city hall did was clearly manipulative and deceptive. The process created a steep learning curve among residents about the politics of development (Keyes, 1969). As such, Boston would have examples of urban renewal that stand as models of how development can work and also projects that stand as monuments to residents' worst fears about renewal and the destructive power of manipulative bureaucratic politics. The hard feelings from renewal policy would be grist for the political mill of a new generation of activists (King, 1981; Lupo, 1981; King, 1986).

While lots of ideas floated around and plans were developed, by the mid-1960s, little had changed for the better in the city. The "new Boston" was just a slogan. More and more whites were leaving, and the last burst of black migration would find large numbers of poor blacks settling in areas the whites had left. The black population would climb from 5% in 1950 to more than 12% in the 1970s.

As Banfield observed, the political role blacks played in Boston had been limited and was largely passive. Unlike in Atlanta, Chicago, or Washington, there were no anchors of black wealth, influence, or access. In the mid-1960s, black banks, large businesses, major institutions, or powerful community-based political actors did not exist. Ed Brooke, a black, was elected U.S. Senator with Republican, suburban Democratic, and independent voters. Black voters, while strong supporters, were hardly more than 5% of the state's electorate. The other major black politician in this period was Tom Atkins. He was elected to one term on the City Council in 1967. Blacks would not score again politically for nearly a decade. Collins would not run for reelection.

In 1967, Kevin White was elected mayor. Like Collins, White was Irish. He defeated Logue, the development director; Louise Day Hicks, the anti-busing activist; and some minor candidates. His election was assured by the strong support of the coalition of business interests favoring development and by black and white middle-class voters who saw him as the most progressive candidate. As mayor, White went on to harness the strong neighborhood sentiments triggered by renewal and to use them as the basis for a network of neighborhood-based public service mechanisms such as "little City Halls," which, as it turned out, were part of a strategy for his political organizing by neighborhood empowerment. The White "organization" kept him in office for 16 turbulent years.

During the early White years, development picked up considerably. Projects on the drawing boards in the late 1960s were finished in the early 1970s. New projects blossomed. Boston's powerful Democratic delegation in Washington delivered funds for projects to the city for everything from antipoverty efforts to Urban Development Action Grants (UDAG) to facility and service improvements. There was a sense of movement and change. White was one in a generation of big-city mayors—Lindsay of New York, Alioto of San Francisco, Cavanaugh of Detroit—whose high-profile leadership in the interest of urban development and liberal urban policies made them players in national Democratic administrations.

Kevin White was viewed as an effective mayor, even as a candidate for higher office. The early politics of busing, however, showed that the suburban progressives, minorities, and developers thought he was a breath of fresh air but that the electoral power still rested with conservative Irish ethnics who never fully trusted this Irishman living in a classy Beacon Hill neighborhood. Mayor White would be called by some of them "Mayor Black."

The politics of busing covers two elections on either side of the 1974 court order on busing. In the election in 1971 (before the busing order was issued), White remained steadfast in pushing liberal politics and a development-oriented city agenda. He encouraged the city to accept desegregation. The School Committee refused. A violent period followed the 1974 order. The mayor stayed aloof as poor blacks and whites fought in the streets and in the schools.

In the elections in 1975 and 1979, White started a steady movement away from any even modest leadership on the city's most critical crisis. Blacks, among others, eventually would blame him for not leading more effectively and constructively. Antibusing forces would be unhappy that he did not embrace their point of view and champion their "alienated rights." There was a clear class struggle. "Poor whites" and their supporters felt under siege from everyone else—blacks, the courts, "limousine liberals" from the suburbs, and "turncoat lace-curtain Irish" (Lupo, 1981; Lukas, 1985).

The mayor did move into even higher gear on his development agenda. Indeed, the success took attention away from other areas. He moved projects that followed on the almost instant success of Fanueil Hall as one of the first of a genre of downtown "festival malls." These buildings came in rapid succession and filled out the skyline that is the present New Boston.

White promoted the glamorous side of urban change, while staying aloof from the day-to-day struggles. Leaderless Bostonians were attempting to write a contract for how to share (or to exclude each other from) the shrinking pie. While new buildings went up and employment increased in steady increments, the best jobs went to suburbanites while Bostonians got the low-paying jobs in hotel laundries and fast-food restaurants. Three quarters of the white-collar jobs went to suburbanites.

While Kevin White did focus on development interests to the exclusion of local conflicts, he also supported reforms. Black leaders and other progressive forces realized that under the at-large system of elections they stood little chance of getting inside city government. On the promise that each neighborhood (district) would get a representative and that there would be four at-large seats, the electorate voted on whether to switch to a combination district/at-large system. The mayor eventually supported the plan and lent the considerable clout of his organization. The switch was approved by voters.

Whereas four all-white neighborhoods used to dominate the City Council and School Committee, power was now more broadly shared. On the Council, blacks won two of the nine district elections. (There were four at-large seats.) Liberal or progressive candidates won at least three others, providing a Council that would, over the last decade, have five to six dependable liberal votes. While not a majority, they would achieve some victories, especially by the mid-1980s when coalitions emerged and a new mayor was in office.

In the late 1970s the liberals took aim at the rich financial gains accruing to downtown developers. Many of the early projects had public incentives and subsidies that some considered overly generous and unnecessary. The call in 1979 from the progressive camp linked downtown development to the still-fledgling development in the neighborhoods by assessing a fee on downtown projects that would then be used to seed neighborhood efforts.

After first opposing this, Mayor White embraced a limited and conservative version of the idea. He appointed a panel to consider the design of a scheme. He would be out of office before "linkage" became a full-fledged program (Keating, 1986).

In the late 1970s, White was linked by the media and the U.S. attorney with corrupt and suspect activities that led to the indictment and jail of more than two dozen of his official and campaign aides. While the mayor himself was never indicted, suspicions mounted. He denied any wrongdoing and left voters wondering until the last minute whether he would crank up the machine that had served him so well in the past. He declined to run for reelection.

In the two years before he left office in 1983, Kevin White became effectively a "lame duck." The nation was in a recession, and problems relating to quality of life for Boston residents continued to worsen. Also by the early 1980s, smoldering issues in various spheres had made their way through the courts. The city's education, tax collection, police hiring, public housing, and other matters were under various degrees of court (or state agency) supervision—from specific remedial orders to receivership. The populace wanted new leadership; and the old Kevin White, who had overcome busing and other obstacles to be reelected in the past, did not have the will to fight again.

In the waning years, various elements of the community did come together to substitute civic leadership (in housing, education, training, and social services) for the failing and discredited administration. These civic initiatives continued to broaden the base of participation, not only by including blacks, but also by adding business people, major employers, state officials, university leaders, and others.

While the mayor did support these civic initiatives and lent some financial support, the administration was a minor player (and at times an obstruction). For example, a group of lenders and community groups joined with the new governor (Dukakis) in forming the Boston Housing Partnership to attack the problem of housing abandonment in the midst of a housing shortage. Some of the city's major employers formed partnerships to support schools and to advance job training and development. Foundations would support both of these efforts, as well as a Human Services Collaborative, to address a host of social problems made more serious by administrative problems, federal cuts, and a state tax limitation initiative.

While Kevin White left the physical shell of a new city, he left a deeply divided populace in which all the new wealth, glamour, and hope seemed to elude ordinary citizens. The lack of leadership meant a transformation without distribution and without any local joy.

THE PROGRESSIVE PERIOD, 1983-1990

The events leading to the election of 1983 included Kevin White's decision process, the jockeying of potential candidates, civic backfilling, court cases, and public forums. All of these were pregnant with the expectation that the election would produce a leader who would end corruption, keep development going, revitalize the neighborhoods, reduce poverty and health problems, bring the races together, and deal with growing federal nonsupport of urban needs. The city, like the rest of the country, was in a serious recession.

In the primary race, there were seven candidates. Five candidates were presented as heirs to the conservative Democratic tradition. A sixth candidate, Mel King, was a black state representative long associated with liberal and progressive causes, including opposition to urban renewal. The final candidate was Ray Flynn, an Irish state representative who had made the political and personal transition from antibusing activist to urban populist.

The primary results put King and Flynn into the two top positions with 60% of the vote between them. It is ironic that a city that had come to be known for conservative Democratic politics and a prodowntown growth orientation would nominate populist and progressive candidates.

After an uncharacteristically civil election campaign, Flynn won the election by an almost 2 to 1 margin. Mel King declared that the "rainbow" had been vindicated in their struggle and had been placed on the political map of the city that so long had shunned them.

Ray Flynn put together an administration that included liberals, blacks, traditional Democrats, and Democratic Socialists in a way no one had expected. No member of Kevin White's senior staff or cabinet had been black. In the Flynn administration, blacks would become city treasurer, public-housing administrator, auditor, school superintendent, election commissioner, as well as heads of other offices. Blacks at the same time increased their representation on the School Committee to 4 of 13 members. A black was elected president of the City Council.

The policies pursued by the administration would be progressive. The administration advocated housing programs, an aggressive linkage program, control of condo conversion, and affirmative action for minorities, women, and gays. The mayor proposed progressive solutions to problems of day care and infant mortality and health, among other issues. He and the governor proposed giving black developers a chance to play the development game in a project that had both a neighborhood site and one in downtown.

Despite these policies, racial friction continued. In 1985, blacks placed on the ballot a petition to incorporate a major portion of the black community as a separate municipality to be called Mandela City. The mayor was opposed to the idea and campaigned actively to defeat it. Few in the black leadership and no one in the larger community leadership embraced the idea. A few black leaders did encourage the issue "to promote discussion about access" or suggest the matter should be studied to see ". . . if it might be feasible." Fewer than 30% of the voters in the affected area voted for the proposal.

Unlike those in other cities, Boston's black communities have not developed strong political institutions. In other cities, these organizations frame the interests of the community, recruit and support candidates, negotiate with other interests, and enforce agreements and solidarity. No such group exists in Boston, and black participation is not part of a consistent strategy. While blacks do respond to such issues as education, they have no clear voice. Attractive black candidates surface in winnable races, but by the time the candidate creates an organization, it is too late.

Boston's political change away from narrow ethnic politics seems permanent. The Yankees slowly have ceded more control to the city. The balance of power (based on demographics and changes in the political structure) has shifted away from the ethnic Irish to a coalition of diverse interests.

In economic terms, Boston has a larger and more diverse labor and business mix, and the physical decay has been greatly reduced. Infrastructure problems exist, but thanks to still-existing restrictions on the city, these are the direct responsibility of the state. The issue is not how far back to ethnic politics the city might recede under a future mayor, but what the speed and direction of reform will be in the future.

REFERENCES

Auger, D. (1979, October). The politics of revitalization in gentrifying neighborhoods. *Journal of the American Planning Association, 45,* 515-22.

Banfield, E. C. (1965). *Big city politics.* Cambridge, Harvard University Press.

Boston Redevelopment Authority (BRA). (1965). *1965/75 General plan for the city of Boston and the regional core.* Boston: Boston Redevelopment Authority.

Boston Redevelopment Authority. (1989). *Boston at mid-decade: Results of the 1985 household survey.* Boston: Boston Redevelopment Authority.

Boston Urban Study Group. (1984). *Who rules Boston: A citizen's guide to reclaiming the city.* Boston: Institute for Democratic Socialism.

Clay, P. (1989a, Fall). Choosing urban futures: The transformation of American cities. *Stanford Journal of Law and Public Affairs 1,* 28-42.

Clay, P. (1989b). *Boston housing: Dynamics, trends, and outlook.* Boston: Boston Redevelopment Authority.

Clay, P., Blackwell, J., Hart, P., Jennings, J., Jones, H., Malson, M., Strickland, E., and Willie, C. (1985). *The emerging black community of Boston.* Boston: University of Massachusetts/Boston Trotter Institute.

Doolittle, F., Jackson, G., Masnick, G., & Clay, P. (1982). *Future Boston: Patterns and perspectives.* Cambridge, MA: MIT Harvard Joint Center for Urban Studies.

Ferguson, R. (1986). *Economic performance and economic development in Massachusetts.* Cambridge, MA: John F. Kennedy School/Harvard University.

Frieden, B. (1989). *Downtown: How America rebuilds cities.* Cambridge, MA: MIT Press.

Gans, H. (1965). *The urban villagers: Group and class in the life of Italian Americans.* New York: Free Press.

Ganz, A. (1987). *The Boston economy and the housing market* (Paper No. 288). Boston: Boston Redevelopment Authority.

Kaufman, J. (1984, November 11). Boston's vital link: The neighborhoods. *The Boston Globe Magazine,* pp. 14-21.

Keating, D. (1986, Spring). Linking downtown development to broader community goals. *Journal of the American Planning Association,* pp. 133-41.

Keyes, L. (1969). *The rehabilitation planning game: A study in the diversity of neighborhoods.* Cambridge, MA: MIT Press.

King, M. (1981). *Chain of change: Struggles for black community development.* Boston: South End Press.

King, M. (1986). *From access to power: Black politics in Boston.* Cambridge, MA: Schenkman.

Kluver, J. (1988). *From home to work: Boston's hotel workers and the prospects for union-sponsored housing.* Unpublished master's thesis, Massachusetts Institute of Technology, Cambridge, MA.

Levin, M. (1960). *The alienated voter: Politics of Boston.* New York: Holt, Rinehart & Winston.

Lukas, A. (1985). *Common ground: A turbulent decade in the lives of three American families.* New York: Random House.

Lupo, A. (1981). *Liberty's chosen home: The politics of violence in Boston.* Boston: Beacon.

Masnick, G. (1989). *U. S. household trends: The 1980s and beyond* (WP W89-1). Cambridge, MA: Harvard University/Joint Center for Housing Studies.

Mollenkopf, J. (1983). *The contested city.* Princeton, NJ: Princeton University Press.

Osterman, P. (1989). *In the midst of plenty.* Boston: Boston Foundation.

Thernstrom, S. (1969). *Poverty, planning, and politics in the new Boston.* New York: Basic Books.

Whitehill, W. M. (1968). *Boston: A topographical history.* Cambridge, MA: Harvard University Press.

Yudis, A. (1984, November 11). $3 billion in projects changing the cityscape. *Boston Globe Magazine,* pp. 22-33.

3

Philadelphia: The Slide Toward Municipal Bankruptcy

CAROLYN TEICH ADAMS

IN THE 1980S, for the first time in decades, observers of Philadelphia's economy glimpsed signs of vigor. The city entered the 1980s well positioned to benefit from the national economic trend toward services, having an even larger percentage of its workers in service jobs than did the nation as a whole. As in Boston, Baltimore, and other Northeastern cities, the new service economy brought thousands of new jobs to Philadelphia and reshaped the downtown skyline, spawning new office towers, hotels and condominiums. Despite this boom in the decade just past, Philadelphia's city government entered the 1990s facing its worst fiscal crisis in 50 years. Philadelphians' average tax bill was among the highest in the nation (Pennsylvania Economy League, 1989). The city had the nation's highest wage tax (about 5%) and the highest transfer tax on real estate (about 4%). Residents paid the highest public transit fares and taxi fares. For three years in a row the city government ended its budget year in deficit.

Recognizing City Hall's fiscal instability, Moody's Investor Service reduced Philadelphia's bond rating two notches to BBB–, the lowest of any U.S. city and the equivalent of rating the city's debt as junk bonds. Only two big cities, Detroit and St. Louis, stood close to Philadelphia's embarrassing position; these two carried the slightly higher bond rating of BBB. Reporting this setback for city officials, *The New York Times* declared "the new urban issue here . . . is survival" (Hinds, 1990, p. A16). To understand how a decade of marked prosperity could end with the municipal government struggling for solvency, we need to trace the recent changes in Philadelphia's civic life that have multiplied the constituencies and responsibilities of local government while eroding its political core.

If one were to stand atop Philadelphia's City Hall and judge the city's recent transition based upon what is visible from the center, one could not fail to be impressed. Large portions of the downtown core and the areas closest to it have been reclaimed from the depressing blight that marked the inner zone in 1960. An ambitious urban-renewal program changed Philadelphia "from a city characteristic of the industrial era to one characteristic of the corporate era" (Kleniewski, 1984, p. 209). Huge sums have been invested by private developers in constructing downtown offices and commercial and residential real estate. Following the rehabilitation of the historic Society Hill district, gentrification transformed several Victorian neighborhoods close to downtown: Fairmount and Spring Garden, near the art museum; Queen Village, adjacent to Society Hill; and Northern Liberties, on the Delaware River. The Market Street East project dominated the redevelopment agenda of the 1970s, resulting in a Gallery Shopping Mall covering several city blocks, complete with a new underground railway station serving the shopping complex. Massive new office construction development west of City Hall in the 1980s shifted the center of downtown activity westward, while on the east side of town the waterfront also underwent drastic physical changes. Several under-used piers were adapted for condominiums, marinas, and restaurants (Bartelt, 1989).

As an office center, Philadelphia prospered, particularly after 1980, on the strength of its business services. Admittedly, this city is no longer a center for corporate headquarters. The clearest example of its decline as a headquarters city is the banking industry. A few decades ago, Philadelphia boasted a dozen large hometown commercial banks. By 1990 it had only one major bank that was still locally-owned and operated, PNB and its parent Core-States. All the others had been merged or taken over by outside interests. Nevertheless, Philadelphia's downtown continued to hold enough corporate activity to support a multitude of downtown firms offering legal, financial, marketing, data processing, and other business services.

Outside the downtown core a different pattern has emerged in the past 30 years. The fabric of entire neighborhoods has unraveled with the exodus of manufacturing jobs from the old industrial districts (Yancey & Ericksen, 1979). At the same time that manufacturing jobs and many manufacturing workers were leaving these sections of the city, blacks were moving into them. Philadelphia's black communities grew from only 13% of the city's population in 1950 to about 40% in 1980. The post-World War II transformation of the region's economy has brought an uneven pattern of decay and redevelopment, increasing the gaps between income groups and generating competition and conflict between races at the lower end of the income scale.

The more demanding educational requirements for jobs in the new service economy impose a disadvantage on youngsters with inadequate schooling. The city has witnessed a dramatic increase in the proportion of young adults, especially men, who are without jobs. For example, by the end of the 1980s, fully three quarters of black men who had not finished high school were out of the workforce; for white male dropouts the proportion was one half (Adams et al., 1991). As labor market conditions changed during the past 30 years, so did family structure. Most notably in Philadelphia, single-parent households increased until in 1988 about 45% of the city's black families and 13% of its white families were headed by single women. Predictably, their chances of living in poverty were far higher than those of married couples. Clearly, the city's changing labor market trends have interacted with racial differences in education and family structure to produce diverging patterns of opportunity for blacks and whites.

Spatial changes accompanied these social changes. The redevelopment of such central areas as Society Hill, Queen Village, and West Philadelphia forced lower income black families out of these increasingly valued neighborhoods and into sections located farther from the center, particularly in North and Northwest Philadelphia. From 1960 to 1990 the degree of racial and economic segregation increased, leading to a greater isolation of poor minorities in parts of the city where the majority of other residents were also poor. In 1980 poor blacks were five times more concentrated than poor whites, and poor Hispanics were eight times more concentrated (Hughes, 1989). These days, it is not uncommon to hear discussions of two Philadelphias—one black and one white.

STALEMATE IN GOVERNMENTAL INSTITUTIONS

A survey of Philadelphia's institutions suggests that very little has changed in the formal structure of municipal government in the last 40 years. The city still operates under the 1951 charter whose reform-minded drafters overhauled several key institutions. They restructured the City Council so that only 10 of its members would represent electoral districts while 7 seats would go to candidates running at-large. To ensure the continual presence of an opposition in the Council, the charter required that at least 2 of the 17 seats must go to the minority party, which since 1951 has been the Republicans. Since 1965, when Banfield reported that the Republicans held 2 seats, their presence has only slightly increased, to 4. Never since 1951 have the Republicans come close to parity with their rivals in this Democratically controlled city.

In keeping with the reformers' ideas of professionalism in city management, the charter created a strong mayor who appoints a managing director to oversee the day-to-day operations of city departments. The charter lodged many of the ancillary functions of municipal government in such independent commissions as the City Planning Commission, the Fairmount Park Commission, the Civil Service Commission, the Historical Commission, and so forth. The city maintains a separate school district overseen by a board whose nine members are appointed by the mayor from names supplied by a citizens' panel. In times of fiscal stress, such as the city now faces, the schools find themselves in competition with city government for tax revenues.

Unlike many other large American cities, Philadelphia's boundaries are contiguous with county boundaries. Philadelphia is both a city and county government at the same time. The city does not get aid from a separate county government that might include some suburban jurisdictions. Furthermore, it is responsible for more than the usual share of services. For example, public welfare, corrections, and health are usually state responsibilities rather than local ones, yet Philadelphia spends heavily on these functions. To supply a broad array of services, the city increased the size of its workforce steadily from 1966 to 1980. Almost every year the city employed more police officers, firefighters, sanitation workers, and others per resident than it had the previous year (Luce & Summers, 1987).

That pattern changed abruptly in the early 1980s, when the city workforce began to shrink. From that moment to the present the almost-constant erosion has reduced the number of employees by almost 25%. Much of this shrinkage was attributable to the cuts in federal funds flowing to Philadelphia, particularly the city's sizable grants under the Comprehensive Employment and Training Act (CETA), which paid workers who swept streets, filled potholes, trimmed trees, cleaned recreation centers, and staffed branch libraries. Like many other cities, Philadelphia used CETA moneys to support basic municipal services. These funds peaked in 1978 during Frank Rizzo's mayoralty, when 4,442 city workers were paid with federal money. By the mid-1980s the Reagan administration had eliminated CETA funds completely. Similarly, the city lost its general revenue sharing funds, which had furnished almost $50 million per year in the late 1970s and early 1980s. Altogether, federal aid to Philadelphia declined from more than $250 million in 1981 to only $54 million in 1990.

To sustain city services during this period of federal cutbacks, the city has relied increasingly on wage taxes, roughly two thirds of which are paid by city residents and another one third by commuters who live in the suburbs but work in the city. The property tax base has supplied a decreasing share of the city's revenues; in fact, the taxable value of real estate per resident has not grown appreciably since 1967 (Luce & Summers, 1987).

As the struggle to maintain services has intensified, so has the tension between the two major branches of municipal government. Banfield's earlier account of the relations between the executive and legislative branches described a strong mayor who, despite his formal powers, had to negotiate with the City Council to achieve his programs (Banfield, 1965). In recent years it would be more appropriate to describe this process as "open conflict" rather than "negotiation." The City Council and the mayor have been increasingly at odds, with the two most recent mayors, William Green and Wilson Goode, championing downtown interests against the Council's more populist perspective.

For example, the City Council successfully resisted a proposal for a major new incinerator to burn trash and recover energy in the form of steam heat—a proposal that was strongly supported by the mayor and by business groups. Faced with a shortage of landfill space and escalating costs to ship the city's trash to other counties and states, both Mayor Green in the early 1980s and his successor, Wilson Goode, sought to persuade the Council to endorse a trash-to-steam plant. The Chamber of Commerce and other business groups emphasized that the impending trash crisis was driving up the cost of doing business in the city and vigorously supported the incinerator as a major priority. Yet the Council resisted intense lobbying by both the mayor and business representatives, siding instead with neighborhood protestors who feared that the incinerator would endanger the public health and their property values. The project was defeated.

Another example of the City Council's resistance to influence by either the mayor or the business lobby was its adoption in the early 1980s of a landmark plant-closing bill that made Philadelphia the nation's first municipality to require firms that close down to give workers two months' notice. As in most cities and states where such legislation has been considered, both the mayor and organized business groups warned that it would scare new businesses away from the city and unduly hamper those already there. The Council was undeterred by the warnings.

In this atmosphere of conflict within the core governmental institutions, Philadelphia increasingly has shifted the planning and execution of such large-scale development projects as convention centers, port improvements, stadiums, and transportation facilities to nonprofit development corporations that are independent of city government. Typically, the financial sponsorship of the projects under their control is shared between the public and private sectors and sometimes involves federal or state funds as well. One reason for their proliferation since 1960 is such independent authorities find it easier to secure financing for projects than does the city government itself—they can borrow money without having the loan count against the municipality's total indebtedness. Moreover, they can insulate development projects from electoral pressure, an advantage that became increasingly important to development advocates as antidevelopment sentiment grew in the 1970s and 1980s. Community activists criticize this practice of putting large-scale development beyond the direct reach of voters because it undermines the democratic character of local government (Friedland, Piven, & Alford, 1984).

RACIAL CLEAVAGES IN THE ELECTORATE

Political historians have emphasized the persistence of political machines in Philadelphia, citing the Republican machine that dominated the first half of the twentieth century and then the Democratic machine built by Congressman William Green in the 1960s. (Congressman Green was the father of Mayor William Green, who served from 1979-1983.) A survey of the last 40 years might at first glance suggest that the Democratic machine remains intact. After all, every mayor during that period has been a Democrat, as have the vast majority of Council members and other officeholders. Almost three quarters of all registered voters in Philadelphia remain Democrats. This apparent hegemony, however, covers a reality of startling disarray in the majority coalition.

A little more than a decade after Democratic reformers swept the Republican machine from power, conflict inside the party was already evident. In 1963 Democratic Mayor James Tate won office without the party's nomination. Tate represented the "rowhouse" Democrats, the working-class white segment of the party, while the reform element controlled the party leadership. Reform leaders were able to deny Tate the Democratic nomination but not the victory in the general election. The reformers found themselves in virtually the same situation when they refused to nominate Tate's police commissioner, Frank Rizzo, to succeed Tate. Rizzo nevertheless successfully campaigned for the mayoralty without the party's endorsement. Rizzo won reelection in 1975, again with the party leadership refusing to endorse him. Some commentators have attributed the reformers' waning influence to suburbanization of their middle-class white constituents. An equally important explanation lies in the reformers' unwillingness to institutionalize their control by using patronage, contracts, and the other spoils of office. They foreswore such trappings of old-style machine politics.

In the process the leadership lost control over nominations—an essential leadership prerogative in any disciplined machine. The party organization could no longer hold a monopoly on instructing Democratic voters. Traditionally, the party organization had distributed "sample ballots" both door-to-door and at the polling places to inform Democratic voters about which candidates had the party's endorsement. In recent elections, however, as many as 100 *different* sample ballots have been printed and distributed by individual Democratic candidates to represent their personal ballot choices rather than the party's. Individual candidates also now distribute the "street money" to election-day workers in wards where they are running—a role formerly performed by the party organization. The power of urban machines depends upon their control over just such mundane aspects of ward politics. Obviously, that control has weakened significantly since the 1960s.

No doubt this centrifugal tendency in the governing coalition is related to the dwindling resources available to officeholders. After all, the classical political machine forged coalitions largely by dispersing patronage and other material rewards for loyalty. As recently as the 1960s, Mayor Tate could still use the patronage and money available through the federal war on poverty to co-opt emerging black leaders to his camp. But the loss of federal funds in the late 1970s and 1980s and the accompanying cutbacks in the city workforce severely restricted the supply of jobs, contracts, and other resources traditionally used to buy political support.

No discussion of the city's changing electoral scene can ignore the growing salience of race as a political touchstone, despite the deceptively nonracial tone of campaign rhetoric. Both black and white candidates in recent contests have publicly eschewed racial themes and racial appeals. For example, Wilson Goode's rhetoric during his 1983 campaign for mayor was resolutely nonracial. He portrayed himself as a reformer and technocrat, pledging a more businesslike approach to managing municipal government and a strong emphasis on such economic development projects as the port and the convention center. He promised to deliver services more efficiently and to create jobs by attracting and retaining businesses. He reminded audiences that there was no black or white way to deliver city programs.

Despite the public disavowals of racial themes by Goode and other candidates, unmistakable signs of a racial cleavage in partisan politics arose. The Democratic Party was increasingly viewed as a black party. In May 1983, when Wilson Goode won the primary to become the Democrats' first black candidate for mayor, about 54% of registered Democrats were white. In 1990, only about 42% were white. A major factor was the 1987 campaign in which Wilson Goode ran for reelection against Frank Rizzo. Rizzo had switched party affiliations from Democrat to Republican to run against the black incumbent mayor, persuading an estimated 40,000 white Democrats to change their affiliation in order to support his candidacy.

As an incumbent seeking reelection, Goode was in an uncomfortable position. The MOVE incident[1] had cost him much of his white liberal support. When Rizzo's candidacy lured many white Democrats to the Republican party, Goode's electoral base narrowed even more. It was clear that he would have to rely more completely on black voters in 1987 than he had in 1983. Recognizing this necessity, Goode's aides worked intensively in the city's black neighborhoods, even while the mayor downplayed race as a factor in his campaign. There was, in short, a tension between the reality of Goode's electoral base (he eventually won by drawing about 97% of the black vote, but only 20% of the white vote) and the rhetorical style of his campaign.

Goode's solid electoral support among black voters should not be taken as a sign that black Philadelphia was totally satisfied with his administration. Like other black mayors in the United States, he surprised his community supporters by his emphasis on promoting economic development and wooing the city's corporate community (Reed, 1988). He successfully resisted a proposal to require minority hiring goals for all government projects built in Philadelphia (Beauregard, 1990), and he clashed often with public employee unions whose membership was largely black. His most visible gesture to the black community was his track record in hiring black administrators for prominent posts. Yet this gesture provided few gains for blacks in neighborhoods. So great was the disillusionment of some black voters that Democratic politicians, both white and black, face in the 1990s the fearful prospect of one of the party's major voting blocks simply withdrawing from the electoral process because of their disappointment with the city's first black mayor.

INTEREST GROUPS IN THE POLITICAL PROCESS

The splintering of the electorate described above is reflected in the tenuous and constantly shifting coalitions that operate in City Hall. For example, despite the Democrats' overwhelming numerical advantage in the City Council, there is no stable legislative coalition. Democratic Council leaders are as likely to seek support from the four Republicans as from their fellow Democrats. We cannot identify stable patterns of interest group activity.

LABOR

Unionized workers furnished a large block of votes for the Democrats who seized power from the Republican machine in the early 1950s. Construction unions were particularly supportive of the reformers' downtown redevelopment agenda. Yet the unions got few rewards from Democratic Party leaders, who recognized that they could appeal directly to workers without relying upon union bosses as intermediaries (Banfield, 1965). A factor limiting union influence has always been its disunity. A recent example is the unions' reaction to the city contract with Marriott Corporation to build a hotel accompanying the new downtown convention center. The Hotel and Restaurant Employees Union, as well as many other labor leaders, opposed

Marriott's bid because of the company's strong antiunion reputation. Yet construction workers supported the choice because of the jobs involved. More powerful than other unions are the public employees. Yet their support for Democratic city administrations has been by no means stable or assured in recent years. Mayors in the 1960s and 1970s secured the support of municipal workers by favorable contracts, often gaining the unions' assent by negotiating generous work rules or pension benefits instead of salaries. For example, Mayor Tate in the 1960s settled a contract dispute with the city's sanitation workers by accepting a clause that enabled the union to block any proposed work-rule changes that would reduce work crews to two persons from the standard crew of three, even after technological changes in trash trucks made it feasible to operate trucks with two-person crews.

The last two mayors, Green and Goode, however, have offered far less generous settlements and thereby drawn criticism and opposition from union leaders. For instance, in order to regain some managerial authority over work rules in July 1986, Mayor Wilson Goode withstood a 20-day trash strike by the city's sanitation workers. Having broken the strike by persuading a judge to issue a back-to-work order, the mayor openly declared his intent to privatize a large proportion of the city's trash collection. The bitter protests that this declaration provoked from city workers demonstrated how quickly a candidate enjoying labor support could become their enemy.

MINORITIES

The emergence of politically independent black factions adds to the complexity of the policy process in city government. Until well into the 1970s, black political participation was brokered by white leaders. Black candidates tended not to make appeals to voters on the basis of race, and the black electorate voted overwhelmingly Democratic whether the party's candidates were black or white. To dilute black influence, the party engaged in gerrymandering of legislative districts (Strange, 1973). Even the numerous protests and racial confrontations that took place in Philadelphia in the mid-1960s did not prompt black Democrats to separate themselves from the party organization. Black political independence finally came in 1978, when Frank Rizzo announced his intention to amend the city's charter so that he could run for an unprecedented third term as mayor. To defeat the referendum on that proposition, black political leaders led a massive voter registration drive in the inner city and enrolled over 62,000 new black voters who turned out solidly to deny Frank Rizzo a chance to run for a third consecutive term. That effort galvanized black voters to start supporting black candidates, including several of the most powerful City Council members of the 1980s and the city's first black mayor.

It could not be said, however, that blacks have operated as an interest group in the sense of sharing an agenda or being organizationally united. We should not expect a population of over 600,000 black residents to share one viewpoint or platform. Black Philadelphia is far too diverse to be represented by a single set of leaders. In recent public debates over the construction of a downtown convention center, for example, some black leaders have favored the project for its economic benefits, while others have opposed it, arguing that it will place an intolerable tax burden on city residents. Debating the city's fiscal crisis, some black leaders support tax increases to maintain services, while others prefer holding the line against tax hikes even if it means cutting services.

Hispanics, who are chiefly of Puerto Rican descent in Philadelphia, represent less than 10% of the city's population and do not yet dominate legislative districts in a way that guarantees them representation in government. The one Hispanic member of the City Council, in fact, was elected at-large rather than to represent a district, which means he cannot concentrate his efforts solely on Hispanic issues. Moreover, Hispanic organizations in Philadelphia are divided into factions, depending upon whether their leaders are native Philadelphians, more recent arrivals from Puerto Rico, or migrants from New York and other U.S. cities.

BUSINESS

During the 1950s a "main line" business elite exercised power through the vehicle of the Greater Philadelphia Movement (GPM), an alliance of civic-minded bankers, lawyers, and business people established in 1949 to clean up the corrupt politics of the Republican machine that had dominated Philadelphia for almost 100 years. The GPM's influence on urban renewal, on the school district, and on a number of other major policy areas was legendary during the reform era (Petshek, 1973) but receded when Mayor Tate restored old-style patronage politics in City Hall in 1963. By actively opposing Tate's candidacy, the business community lost much of its influence on municipal government. During Mayor Rizzo's two terms as well, the business community remained at arm's length from City Hall, a situation that led business leaders to contribute heavily to defeating Rizzo's bid to change the charter—a change that would have left him free to run for a third term.

Business leaders tried in the early 1980s to remobilize themselves to regain access to City Hall by resurrecting the Greater Philadelphia Movement in a new form, renamed the Greater Philadelphia First Corporation (GPFC). The organizers recruited the chief executive officers of the region's 27 largest corporations to become members of the board of GPFC. Each of their companies agreed to contribute a minimum of $50,000 annually to create a pool of funds that pays a professional staff and finances this ambitious group's projects. One of the first opportunities presented to this revitalized business coalition came in the 1983 mayoral campaign, when business leaders enthusiastically backed Wilson Goode. They were impressed by his previous record as the city's managing director and by his willingness to seek their advice during the campaign. Once elected, the new mayor asked for, and accepted, guidance from business supporters in choosing his cabinet officers, department heads, and other key appointments. This was the most direct influence over municipal administration that business people had exercised since the heydey of the reform movement in the 1950s.

Despite this burst of energy in the early 1980s, the reincarnated business coalition made a disappointing showing. Its efforts were marked by stagnation. It seemed unable to achieve any of its goals. Most disappointing was the GPFC's failure to advance two of the largest public-works projects proposed in the 1980s—the trash-to-steam incinerator mentioned earlier and a downtown convention center. The incinerator plan is dead and the convention center has been stalled for years, testifying to the weakness of business power. The track record of the GPFC, said one of the city's foremost business columnists, "made an oxymoron of the term 'business leadership' " (Binzen, 1990, p. E3).

The extent of business frustration was evident in a recent article in *Philadelphia Magazine,* titled plaintively, "Why Can't We Be Like Cleveland?" The article lauded the role of business leaders in Cleveland who operated as "gutsy, united CEOs" (Huler, 1989, p. 99) to reshape municipal priorities through a group known as Cleveland Tomorrow. It reported that in the summer of 1989 two dozen of Philadelphia's top business executives had met behind closed doors to look at Cleveland as a model of how businesses can mobilize to save a declining city. Yet in a city so balkanized as Philadelphia in the 1990s, it is hard to imagine business reasserting its vaunted influence of the 1950s, particularly since so many CEOs are transients sent into the city to manage branch offices of companies whose headquarters are elsewhere. Their stake in city politics is simply not as great as that of corporate heads 40 years ago.

CITY AND SUBURBS: CHANGING RELATIONSHIPS

Like other Northeastern and Midwestern urban centers, Philadelphia no longer dominates the surrounding region either politically or economically.

Beyond the city's boundaries, which essentially are unchanged since 1854, seven suburban counties in Pennsylvania and New Jersey account for much of the economic growth and virtually all of the population growth in the region in recent decades. Job growth in the suburbs since 1980 has been different from that in the central city. While the city has spawned jobs at either very low or very high wages, the suburbs have generated jobs in all wage categories, from entry level to managerial, including large numbers of middle-income jobs (Adams et al., 1991). That is, the suburbs have maintained an economic base for the middle class that the city has been losing. Not only the socioeconomic profile of suburbanites, but also their political profile, differs from the central city. While city dwellers are overwhelmingly registered as Democrats, Republican registrations predominate in the surrounding counties by ratios of 2 to 1 and 3 to 1.

The single biggest cleavage dividing political leaders from the city and suburbs, however, is not partisan ideology, but a bread-and-butter issue—Philadelphia's wage tax. Since World War II the city has levied the wage tax on all workers who held jobs in the city, regardless of where they lived. Suburban politicians always have bridled at what they regarded as taxation without representation. Their constituents who commute to work in the city have had no voice in electing the City Council that sets tax rates. Suburban politicians' continual chafing at this perceived inequity has colored their responses to many policy issues. City and suburban cooperation on the development of a regional port facility, the maintenance of the regional transit system, the upgrading of the public education system, and many other issues crucial to the development of the region has been hampered by the festering disagreement over the wage tax. The feud spills over frequently into the state legislature, where delegations from the city and the suburban counties cannot agree even on crucial projects and programs. For example, in the spring of 1986 several dozen state legislators from communities on Philadelphia's border tried to block a state subsidy for the construction of the city's proposed convention center. Despite the then-Republican governor's contention that the project would be a boon for all of southeastern Pennsylvania, these suburban Republicans voted against it. In doing so, they openly admitted that their opposition was a tactic designed to force concessions on the wage tax. While they were ultimately unable to defeat the appropriation to Philadelphia, their delaying tactics added to the general climate of suspicion that has marked political relations between the inner and outer rings of the region.

The devolution of responsibility from the federal to state level in the past decade has made the legislature and governor more important players. City officials must cultivate their support yet sometimes have been stymied in doing so, as the convention center example shows. Philadelphia's power at the state level is waning, partly because its delegation is shrinking. As late as the 1970s, Philadelphia had 5 more House members and 2 more senators than it does now. With reapportionment after the 1990 Census, the city's numbers will shrink more. Even the 29 House members and 7 senators who continue to represent the city often do not present a united front.

In the case of the convention center, for instance, the delays in state funding and state approval for the quasi-public authority to manage the project were attributable not only to resistance from suburbanites outside the city limits. Equally important was the disunity within the city's delegation. Some of the city's black legislators refused to support the project without an assurance of minority set-asides—a position not shared by the city's black mayor, Wilson Goode. When the city finally did succeed in getting state legislation passed to allocate money to the city's project and to set up a new Convention Center Authority, almost half the language in the bill dealt with ethical restraints on local politicians, prohibiting the hiring of relatives or the letting of nonbid contracts. These restraints reflect the state legislature's view of the city's political processes as contaminated by corruption, parochialism, and incompetence.

Not only major projects like the convention center, but also everyday needs within the city, go unmet because of the state government's stance toward its largest urban center. Legislators in Harrisburg see Philadelphia as the city that has only 14% of the state's population but 40% of Pennsylvania's public-assistance recipients and 50% of its AIDS victims, while producing about 50% of the prisoners in the state correctional system. They traditionally have underfunded many of the social services that they rely upon the municipal government to provide. Philadelphia ranks behind other major cities in the Boston-Washington corridor on the measure of general per capita state aid, as well as unrestricted state aid (PEL, 1989). Comparing Philadelphia with Boston, the other Northeastern city included in this volume, one sees that Philadelphia relies far more heavily on locally generated revenue; the ratio of local taxes to state aid per capita is 4 to 1 in Philadelphia, but only .7 to 1 in Boston (PEL, 1989). Boston, New York, Newark, and Baltimore get more revenue from state legislatures than they raise locally, while Philadelphia receives far less. In numerous policy areas, particularly services to children and youth, services to mentally retarded, and services to mentally ill homeless people, the city has been "overmatching" state aid in recent years—spending more than its statutory share while getting less than the state's statutory share of the cost of these services. Especially galling to city officials are service areas in which the state mandates specific standards of service but then underpays for them.

POLITICAL REGIME CHANGE

Clarence Stone's notion of urban regimes connotes dominant coalitions that can sustain themselves over years, even decades. I have argued that there is no longer any such sustained coalition operating in Philadelphia. The pattern of political change since the 1960s has been increasingly centrifugal, creating a Democratic Party that is nominally in control yet without a center. Perhaps this is not quite what Yates (1978) had in mind when he coined the term *street-fighting pluralism*, for he was depicting the politics of street demonstrations, community activism, and open neighborhood protest that marked the 1960s and 1970s. That atmosphere of open confrontation and bargaining over neighborhood interests existed briefly in Philadelphia during the late 1970s, but it was replaced during the 1980s by a revival of ward politics led by neighborhood-based politicians who operated inside the state and municipal governments rather than outside those formal structures. The real community leaders in Philadelphia today are either ward leaders and officeholders at the local, state, and federal level, or they are directors of nonprofit development corporations who are more likely to spend their time negotiating loans than organizing protests.

If the city does not exhibit the "street-fighting" character Yates was describing, it nevertheless conforms to his notion of extreme pluralism of political, administrative, and community interests that are "fragmented to the point of chaos" (Yates, 1978, p. 34). The City Council and the mayors of the 1980s have been constantly at odds. Journalists have tended to attribute Mayor Goode's weakness to his loss of prestige in the MOVE disaster. But the fact that his predecessor was in a similar situation vis-à-vis the City Council argues for a more general, less particularistic interpretation. Without a reliable support coalition in the City Council, neither mayor could pursue his policies effectively.

Much of the literature on urban regimes has focused upon organized business as the single most powerful, sustained influence in urban politics. What Stone calls "corporate" or "entrepreneurial" coalitions are seen to have their hands on the levers of power in many cities. Yet in Philadelphia, business interests have found it increasingly difficult over the past 30 years to advance their agenda because they cannot find the levers to grasp. They have tried, to a great extent, to circumvent the political realm by adopting privatist solutions to economic and social problems.

An example is the corporate response to inadequacies in the region's transportation system. Consider a typical suburban transit problem. King of Prussia is a burgeoning office and retail node in suburban Montgomery County whose growth has created thousands of entry-level jobs. Yet entry-level workers cannot find affordable housing near their jobs in this affluent area. A 1989 demographic study commissioned by the suburban Chamber of Commerce concluded that more than half the households living within 3 miles of this development have estimated annual incomes of $50,000 or more. Almost a third have incomes over $75,000. To find low-wage workers, suburban employers must go very far afield, including the inner city. Unable to rely upon the public transit system to handle the reverse commuters they have lured to work in the suburbs, suburban corporations in King of Prussia and elsewhere have created privately-financed shuttle bus services.

Another example of this "self-help" approach is the downtown business response to inadequate municipal services in the central city. In 1988 one of the city's largest department store owners, G. Stockton Strawbridge, organized the retailers occupying the redeveloped blocks of Market Street East to spend their own money for a private cleaning and security force to supplement the municipal services in their area. Taking that model a step farther, a downtown developer in 1990 pushed the city government to approve a special-services district comprising 80 downtown blocks, similar to those operating in New York, Denver, Cleveland, and New Orleans. Business owners in the downtown business core pay an extra 4.5% on their property taxes to fund extra police patrols and more street cleaning. A 21-member board of directors, chaired by the developer who initiated the idea, oversees the district. In addition to such basic housekeeping services as transportation, sanitation, and security, some observers in the region even see private businesses playing an expanding role in helping employees obtain housing, education, health care, child care, and legal services as federal and state programs decline (Schwartz, Ferlauto, & Hoffman, 1988).

I have portrayed corporate power as inadequate to move city government in the 1980s and probably the 1990s as well. Does business weakness afford other interest groups in the community more opportunity to influence policy than in cities dominated by business coalitions? Probably not. The same fragmentation that frustrates business lobbyists frequently defeats the efforts of community and public-interest activists as well. Consumer advocates, environmentalists, tenants' rights organizers, and other progressive political activists find it easy to gain a sympathetic hearing from one or more leaders in local government. But the hyperpluralism makes it impossible to forge effective alliances to push for policy change.

Banfield's earlier account predicted that in Philadelphia "eventually reform is likely to return as a dark horse" (Banfield, 1965, p. 107). Banfield's optimism is unjustified by any change to date. The one glimmer of a reform impulse backed by significant money and power is the coalition now emerging around the leadership of Congressman William Gray, a black minister from North Philadelphia who holds the position of majority whip in the U.S. House of Representatives. For several years Gray has worked to establish himself as the center of a reform coalition by supporting his handpicked candidates for City Council, for the State Senate, and for mayor in 1991. He has lavished campaign contributions upon his allies, in addition to using his considerable influence in the city to promote their candidacies. Gray has pushed hard for a multiracial coalition, being careful to support white liberals, as well as progressive blacks. His candidates have not always won. For example, his white candidate for district attorney in 1989 failed to beat a popular Republican incumbent despite gaining the support of 85% of black voters. But he has helped build an incipient coalition of four reform Democrats in the City Council who have begun to operate collectively. He can take credit for enough wins to be a major player in selecting the next mayoral nominee of the Democratic Party in the city.

CONCLUSION

I have tried to analyze how a decade of marked prosperity could end with the municipal government struggling for solvency. These days this question is usually formulated in public debates as a technical issue involving the trade-offs among different types of taxation, different subsidy formulas, and their differential impacts on businesses and homeowners. Poor management is cited commonly to explain the city's fiscal problems. But the real explanation is political and hinges on the fragmentation that has been the focus of this chapter. None of the cures for Philadelphia's failing fiscal health can be applied without political consensus. The city cannot hope to raise taxes, cut services significantly, or get more money from the state and federal governments without overcoming the divisions and infighting that mark its majority coalition. Having no coordinated framework for bargaining and trade-offs, city officials find it impossible to make the tough choices and apply the bitter medicine of higher taxes or service cuts. City officials cannot effectively target the scarce resources they *do* have in order to make an impact on urban problems. Instead, the city's resources must be apportioned to accommodate the multiple power centers. One documented example is the insistence by City Council members on building new playgrounds, swimming pools, recreation centers, and other community facilities in their own districts when budgetary shortages make it impossible to keep the existing facilities open and staffed (Adams, 1988). The ultimate lesson contained in this analysis is that no technical fix will restore Philadelphia's fiscal and civic health. If the essential problem is political in nature, so is its solution.

CAROLYN TEICH ADAMS   45

NOTE

1. In a predominantly black neighborhood in West Philadelphia in May 1985, community residents were threatened with violence by members of MOVE, a small commune of black extremists who preached a back-to-nature philosophy. Urged on by local residents, the Philadelphia police tried to force MOVE members to abandon their fortress-style house. The stand-off between police and commune members turned to tragedy when a police helicopter dropped explosives onto the roof of the MOVE house, setting off a fire that killed all 11 MOVE members barricaded inside the house and burned down two city blocks. Wilson Goode's mismanagement of the emergency led many white liberals to condemn the use of excessive force to quell a neighborhood disturbance, and many black supporters of the mayor were similarly critical of the police over-reaction.

REFERENCES

Adams, C. (1988). *The politics of capital investment: The case of Philadelphia*. Albany: State University of New York Press.
Adams, C., Bartelt, D., Elesh, D., Goldstein, I., Kleniewski, N., & Yancey, W. (1991). *Philadelphia: Neighborhoods, division and conflict in a post-industrial city*. Philadelphia: Temple University Press.
Banfield, E. C. (1965). *Big city politics*. New York: Random House.
Bartelt, D. (1989). Renewing center city Philadelphia: Whose city? Which public's interest? In G. Squires (Ed.), *Unequal partnerships: The political economy of urban redevelopment in postwar America* (pp. 80-102). New Brunswick, NJ: Rutgers University Press.
Beauregard, R. (1990). Tenacious inequalities: Politics and race in Philadelphia. *Urban Affairs Quarterly, 25*, 420-434.
Binzen, P. (1990, January 1). For Philadelphia, ten turbulent and productive years. *Philadelphia Inquirer*, p. E3.
Friedland, R., Piven, F. F., & Alford, R. A. (1984). Political conflict, urban structure, and the fiscal crisis. In W. Tabb & L. Sawers (Eds.), *Marxism and the metropolis* (pp. 273-297). New York: Oxford University Press.
Hinds, M. D. (1990, June 21). After renaissance of the 70s and 80s, Philadelphia is struggling to survive. *New York Times*, p. A16.
Hughes, M. A. (1989). *Poverty in cities*. Washington, DC: National League of Cities.
Huler, S. (1989, October). Why can't we be like Cleveland? *Philadelphia Magazine*, pp. 99-107.
Kleniewski, N. (1984). From industrial to corporate city: The role of urban renewal. In W. Tabb & L. Sawers (Eds.), *Marxism and the metropolis*. (pp. 205-222). New York: Oxford University Press.
Luce, T. F., & Summers, A. A. (1987). *Local fiscal issues in the Philadelphia metropolitan area*. Philadelphia: University of Pennsylvania Press.
Pennsylvania Economy League (1989, June). *Revenue and expenditure comparisons: Philadelphia vs. selected major cities*. Philadelphia: Author.
Petshek, K. (1973). *The challenge of urban reform: Politics and programs in Philadelphia*. Philadelphia: Temple University Press.
Schwartz, D. C., Ferlauto, R. C., & Hoffman, D. N. (1988). *A new housing policy for America*. Philadelphia: Temple University Press.
Strange, J. H. (1973). Blacks and Philadelphia politics: 1953-1966. In M. Ershkowitz & J. Zikmund (Eds.), *Black politics in Philadelphia* (pp. 109-144). New York: Basic Books.

Yancey, W. L., & Ericksen, E. P. (1979). The antecedents of community: The economic and institutional structure of urban neighborhoods. *American Sociological Review, 44,* 253-262.
Yates, D. (1978). *The ungovernable city.* Cambridge, MA: MIT Press.

4

Chicago: Power, Race, and Reform

BARBARA FERMAN

SINCE GOSNELL'S CLASSIC STUDY in 1937 of machine politics, Chicago has been the laboratory for numerous studies[1] on that "institution peculiar to American politics" (Lowi, 1968, p. v). The Daley administration provided further grist for the empirical mill of machine studies by achieving a level of centralized power rare for urban politics.[2] Between 1960 and 1990, however, Chicago experienced major economic, demographic, and structural changes. The political and policy realignments that resulted make for an even better study in regime change.

DEMOGRAPHIC AND ECONOMIC CHANGE

Chicago's population declined from 3,550,404 in 1960 to 2,784,000 in 1990, forcing it to cede its famed "second city" status to Los Angeles. Behind the aggregate numbers lay major racial and economic change: Chicago's black population increased from 22.9% in 1960 to 41% in 1985; the Hispanic population nearly doubled in size between 1970 and 1980; and the 1980s have seen the growth of Chicago's newest immigrant group, Asians. Census estimates for 1990 show a population that is 42% black, 36% white, 18% Hispanic, and 4% Asian (McCarron, 1990). Economically, the trend has been downward: Between 1959 and 1979, Chicago lost 35% of its middle- and 37% of its upper-income families; at the same time, the number of low-income families increased by 40.2% (Squires, Bennett, McCourt, & Nyden, 1987).

AUTHOR'S NOTE: An earlier and more detailed version of this chapter was presented at the 1990 Urban Affairs Association Meeting as "Chicago: Political and Administrative Change, 1960-1990." Special thanks to William Grimshaw for comments on that draft.

Deindustrialization combined with migration patterns to further depress economic conditions. Long heralded as the industrial capital of the Midwest, Chicago began losing manufacturing jobs after World War II, with the most precipitous declines occurring between 1972 and 1983 (40.7%). And, for the first time, the suburbs showed losses (17.9%) (Weiss & Metzger, 1989).

The city's response to economic decline was a progrowth strategy of downtown office construction and selective neighborhood revitalization. As a result, service sector employment surged (from 17% to 27% of total employment, 1950-1980) as did downtown real estate values. Neglected by this strategy were the displaced workers and older neighborhoods where industrial decline took the first toll and disinvestment and redlining completed the process of deterioration. Consequently, Chicago has a glistening downtown surrounded by expensive residential towers, while vast tracts on the South and West Sides reveal the social and economic costs of uneven development; high unemployment, excessive poverty, and physical abandonment have replaced the factories that fueled the industrial heartland.

While these neighborhoods are almost totally black, white working-class neighborhoods also bore the brunt of disinvestment. The deterioration is not as great, but these neighborhoods are further evidence of a strategy that clearly delineated between the downtown and the neighborhoods.

THE METROPOLITAN CONTEXT

Chicago and its suburbs for a long time constituted a study in contrasts. The former was Democratic and practiced that raucous brand of politics typical of urban machines. The latter were predominantly Republican and featured uneventful versions of reform politics. Chicago was Carl Sandburg's city of "broad shoulders," while the suburbs housed William Whyte's "organization man."

Through electoral power, political domination of the Cook County Board of Commissioners (which controls vital county offices), and the machine's clout in county and statewide politics (Republican as well as Democratic), Chicago was able to ignore its suburban neighbors without much consequence.

Population and employment shifts, however, altered the political balance of power and the nature of policy debate. As Chicago's share of the county population fell from 80% in 1950 to less than 60% in 1988, its vote in countywide elections was similarly diluted. While this decline helped reconfigure city-suburban relations, the suburbs are not a monolithic bloc. Indeed, they share with Chicago the phenomenon of uneven development; some suburbs have prospered economically, while others resemble Chicago's "inner zones" where manufacturing and population losses have taken a devastating toll. This variation has resulted in different patterns of city-suburban relations.

One pattern finds Chicago engaged in corporate location battles with the growth suburbs. The major weapons are various tax incentive packages. State development policies also become part of the battle. Here Chicago is joined by the poorer suburbs that cannot compete in the location battles but can certainly ally with the major dissenter for a bigger share of state largesse.

Tax issues tend to reinforce the city/growth-suburb cleavage; the city favors an increased income tax and property tax relief, while many suburbs support a reduced income tax. Complicating the debate is the larger antitax movement, which is a major factor in state electoral politics.

Political cleavages are developing also within and between the growth suburbs as the results of unplanned and rapid development threaten the physical, social, and economic fabric of these communities. Concerns over infrastructure and quality-of-life issues have spawned numerous limited-growth proposals. These measures, however, exist alongside a fierce competition between the suburbs to attract business.

GOVERNING CHICAGO: A STUDY IN REGIME CHANGE

Chicago's political arrangements and policy orientation have undergone significant change over the last three decades. A progrowth regime, based upon a narrow governing coalition of "selective insiders" and supported by a machine style of politics, was supplanted by a broad-based progressive coalition characterized by a reformist political orientation. This regime transition was fueled largely by fundamental shifts in the organizational base of electoral politics and a declining stock of discretionary resources, two features that became mutually reinforcing.

The progrowth regime's emphasis on downtown development fostered neighborhood opposition, while the political machine's insensitivity to blacks stimulated racial discontent. These twin rallying points shifted the dynamics of electoral politics. As electoral mobilization increasingly took on a group dimension, the nature of demands shifted from individual to collective benefits, thus requiring a new accommodation process. At the same time, the machine's discretionary resources (patronage and selective incentives), which helped maintain a politics of the individual, were in decline. These shifts created serious challenges to the composition and policy orientation of the progrowth regime.

Ultimately, these challenges resulted in the electoral victory of a progressive reform coalition. Through appeals to collective, rather than individual, benefits, the coalition brought together blacks, Hispanics, white liberals, and activist neighborhood organizations. The emphasis on collective benefits put policy at the center of the accommodation process, while the breadth of the coalition ensured that the policy would be much more inclusionary than that of the progrowth regime. Finally, the increased importance of policy expanded the role of city government from facilitator of downtown growth to promoter of balanced development and fairness.

THE DALEY REGIME: MACHINE POLITICS, PROGROWTH POLICY

Chicago's progrowth regime was spearheaded by Richard J. Daley and based on a narrow coalition of "insiders" from the business and labor communities. The policy focus was almost exclusively downtown development, with planning and implementation activities carried out largely by these insiders from the business community. City government attended primarily to basic housekeeping functions, while top-down redistributive programs came via federal largesse and were at the bottom of the priorities list.

The regime was sustained by an accommodation process that featured a separation of electoral and policy politics. By structuring the electoral arena around the distribution of material rewards and incentives, Daley kept electoral and neighborhood politics on an individual and issueless basis. Discretionary resources, in the form of patronage and "insider privileges," provided a buffer between economic elites and voters, thereby allowing Daley to deliver policy benefits to those economic elites.

The dynamics of the accommodation process are particularly evident in the case of organized labor and blacks, two groups that stimulated significant conflict in other cities but that were relatively passive in Chicago. While Daley was a "labor man," he did not endorse collective bargaining rights for unions. Rather than preside over the development of strong organizations, Daley kept labor happy with high salaries and labor leaders with good appointments.[3] Under the prevailing wage system, union workers in Chicago were among the highest paid in the nation. A thriving economy, combined with Daley's progrowth orientation, translated into many construction jobs. Monetarily, organized labor did well by Daley. Consequently, Chicago enjoyed labor peace, while cities like New York were wracked by labor turmoil.

A similar strategy was applied to blacks. Material resources were used to co-opt black leaders, while policy issues of concern to the black community were ignored. This strategy of separating leaders from the led prevented the former from developing any independent organizational base and thus ensured their loyalty to, or at least dependence upon, the machine. Black voters were also dutiful supporters of the machine, giving Daley his edge of victory in 1955 and 1963. Ultimately, this strategy collapsed as growing problems in the black community, combined with the machine's insensitivity, resulted in massive antimachine mobilization.

The policy arena under Daley consisted largely of economic-development strategies. When Daley took office, Chicago's economy was in decline. Manufacturing was being lost to the suburbs, while the central business district (CBD) was suffering physical blight and falling property values. As in many cities, a progrowth strategy, centered around economic restructuring and corporate expansion, was adopted. The development emphasis was on large-scale office construction and infrastructure improvements in the CBD, as well as selective neighborhood renewal.

The division of labor between the private and public sectors resembled that between leading actor and support cast. The plans were developed and carried out by the Chicago Central Area Committee (CCAC), a group of key business leaders designated by Daley as the "representatives" of their community. As such, they had special access to city government. The path was simplified with the establishment of a Planning Department in 1957 to serve as the instrument of downtown development. Given the fragmentation of Chicago city government, the establishment of one agency was crucial. Further, Daley's planners shared the progrowth philosophy of the CCAC.

Equally important was the city's capital investment strategy. While much of the construction was privately financed, capital investments were used for site acquisition and preparation and for infrastructure improvements. In addition, the construction of large government complexes served as important anchors. Not surprisingly, capital investments were concentrated in the downtown, O'Hare International Airport, and Near North Side areas (Greer, 1986). Thus, the policy orientation and capital investment strategies, in dividing the city between downtown and the neighborhoods, mirrored the political dichotomy between policy (economic development) and electoral (neighborhoods) politics.

Outside of economic development, city government assumed a caretaker role, providing the basic housekeeping functions: police, fire, sanitation, education, parks, and public works. This routine service, rather than policy, orientation partly reflected the machine's conservative philosophy: Such nongovernmental entities as the family, church, neighborhood, and ward organization could better serve the individual than could government programs. It also reflected the fiscal conservativism of the machine: Limiting government to basic housekeeping kept expenditures down. Finally, it reflected the pragmatic politics of the machine: Services allocated in the "friends and enemies" tradition of machine politics enhanced the supply of discretionary resources, while maintaining an issueless and individual politics. Even federal programs were subjected to this pragmatism, as legislative intent was usually subordinated to local political needs.

CHALLENGING THE REGIME

Maintaining Chicago's growth regime required a continuing capacity to centralize power in the political and governmental arenas. A combination of challenges from within the political machine, a decrease in discretionary resources, and electoral shifts weakened that capacity. Although the most visible signs appeared after Daley's death in 1976, there is evidence of a rupture in the accommodation process during his administration.

Politically, 1972 was a watershed year in Chicago. A slate of independent candidates, drawn from the Republican and Democratic Parties and combining extensive media usage with independent campaigning, won numerous offices, including that of governor. Black Congressman Ralph Metcalfe broke with his lifelong mentor—Richard J. Daley—over the issue of police brutality in the black community. Ed Vrdolyak, City Councilman and vocal spokesman for the Young Turks on the City Council, convened a series of secret meetings with other aldermen who wanted more say in legislative matters. Nationally, the Democratic Party sent a strong message when they refused to seat the Daley delegation for its failure to comply with the newly enacted McGovern rules. In the courts, political firing was declared unconstitutional in the first of the Shakman Decrees.[4]

The Daley machine thus suffered serious challenges on several fronts: its ability to win elections, its control over the black vote, its control over its own members in the city council, its traditional "special status" within the national Democratic Party, and its control over patronage. Although Daley won another election, albeit with sharp vote reductions, these events signaled a changed political landscape.

The most fundamental change was in the organization of the electorate. The issueless and individual politics of the machine were being replaced by controversy and group mobilization. The machine's continued insensitivity to the growing problems in the black community stimulated black mobilization. Economic downturn raised challenges to the progrowth agenda and became a rallying point for neighborhood mobilization. Thus, a sharp split was developing between the nature of electoral demands and the response of the governing coalition.

RACIAL POLITICS

This split was particularly evident in the governing regime's relationship to the black community. As early as the mid-1960s, there were signs that race was becoming the foundation of Chicago politics.[5] White middle-class voters increased their support for the machine, while blacks, who had been the most loyal supporters, began withholding their support. The first development is unusual for machine politics, which appeals to voters of lower socioeconomic status through the extension of material incentives. It was not unusual, however, for the new direction in Chicago politics; middle-class whites began to see the machine as the defender of white racial interests.

The machine happily obliged. By appointing anti-integrationists to the Chicago Housing Authority and Board of Education, two institutions pivotal in the battle for racial equality, Daley sent a clear message to white voters. The message was even louder in his refusal to negotiate the police brutality issue, and near deafening in his "shoot to kill" orders issued during the riots following Martin Luther King's assassination.

The "circularity of politics and policy" that Stone (1989) attributes to a regime's accommodation process was undermining the Daley regime. While blacks had yet to find an alternative to Daley, they had begun to withhold their support. This development, although downplayed by the machine, foreshadowed a more serious development—the split between black machine politicians (elites) and black voters (masses). While the former did benefit from supporting the machine, the latter received little to no benefit. Thus, they had little to lose by breaking with the machine. The "selective incentives" and "insider privileges" (Stone, 1989) that are extended to elites did not apply to the masses. Nevertheless, the governing regime was not about to address the collective issues of race and discrimination, lest it risk its white support.

The machine continued to take the black vote for granted and, in doing so, stimulated black political mobilization. The selection of Michael Bilandic to finish out Daley's term was the first in a long series of actions that angered and mobilized the black community. While the rules of succession were unclear, many in the black community felt that black City Councilman Wilson Frost, as President Pro Tem of that body, was the rightful heir. Although Frost was given the powerful position of Finance Committee Chair, it was seen as another slap in the face to the black community.

The last slap that the black community took from the machine came during the blizzards of 1978-1979. To expedite mass transit rush-hour service to outlying (white) areas, the Transit Authority bypassed several inner-city (black) stations. The black community retaliated at the polls when they helped Jane Byrne defeat Michael Bilandic and, hence, the machine. While Bilandic had the support of black elected leaders, Byrne had the voters, winning 14 of the 16 predominantly black wards (Green, 1987).

This split between black leaders and black voters signified a major shift in Chicago's electoral politics, foreshadowing the need for a new accommodation process. Blacks were identifying as a group and demanding collective benefits, while the old regime still was responding with particularistic benefits.

Byrne won the battle by exploiting this contradiction but lost the war when she too ignored the new political reality. Through personnel and labor actions, which included a reneging on campaign promises, Byrne alienated the black community early on. Like Daley, Byrne used housing and school board appointments to shore up her white electoral support.

The alienation deepened during a series of strikes in the winter of 1979-1980. Unlike Daley, Byrne took a strong stand against the strikers, even threatening the use of court injunctions. Organizationally and fiscally, however, Byrne was operating in a different environment. The unions, particularly teachers and transit workers, had become more radicalized. Fiscally, the problems included Board of Education and city deficits ($101 million and $102 million, respectively) (Holli, 1987). Thus, Byrne had union representatives who demanded and expected more, and she had less to give. Nevertheless, her position was another straw on the bad back of race relations, since the services being struck (transit, fire, and education) disproportionately affected blacks.

Once again, the black community registered its discontent at the polls, contributing to major electoral defeats for Byrne's slate in the 1980 Illinois primary races. The real electoral prize, however, came in the mayoral primary of 1983, when Jane Byrne and Richard M. Daley (son of Richard J. Daley) split the white vote, thereby paving the way for the city's first black mayor— Harold Washington.

NEIGHBORHOOD POLITICS

Although Chicago is home to the most renowned community organizer, Saul Alinsky, community organizations faced a tough time. Perceived as threats to the ward organizations and thus to the machine, they were denied access to local resources and to decision-making bodies. This began to change in the late 1970s, though, as the increase in media campaigns and the decrease in patronage weakened the ward organizations. Moreover, the fallout from economic restructuring helped galvanize communities around the major policy issues of neighborhood development, job retention and creation, affordable housing, and education. Community-based organizations (CBOs) were the logical beneficiaries of this shift in neighborhood politics and priorities.

Jane Byrne was among the first candidates to recognize the electoral possibilities. Supplementing her appeals to the black community were major overtures to the neighborhoods, declaring them to be her first priority. Setting up a Department of Housing and a Department of Neighborhoods, Byrne gave institutional recognition to her campaign promises.

Unfortunately, these overtures suffered the same fate as her appeals to the black community and for the same reasons. Byrne was elected with a coalition of blacks, lakefront liberals, and disaffected white ethnics. She could neither turn this into a governing coalition nor maintain it as an electoral coalition; appeals to one part would alienate the others. Moreover, the first two groups had no influence within the machine or within city government, while the third group consisted mainly of Northwest Side Polish. The Polish had always played second fiddle to the Irish in Chicago politics. Finally, Richard M. Daley was emerging as a serious rival in the upcoming mayoral election.

Byrne responded to the changing political landscape by making peace with the very machine she ran against. That treaty spelled the end of collective appeals to the black community and the end of any substantive neighborhood policies. The offices she set up went the way of machine politics, becoming funnels for patronage.

RESTORING THE GROWTH MACHINE

Driven by electoral concerns, Byrne embarked on a progrowth strategy in an attempt to appeal to the boosterism of the major newspapers and to attract campaign support from the development community. In contrast to Daley, however, Byrne could not centralize enough power to deliver the necessary political and administrative support. She also faced a more organized opposition. Finally, Byrne shifted the role of city government from supporter of economic development to direct promoter, a much costlier approach and one that further solidified the opposition.

Whereas Daley had operated under the philosophy that business should finance its own development, Byrne came armed with tax abatements, industrial revenue bonds, and UDAGs as she proposed commercial, residential, and tourism projects for the downtown. The capstone of this development fever would be the 1992 World's Fair, which promised to rearrange land use patterns completely on the south edge of the CBD to the tune of nearly $1 billion. While the level of public subsidy was unclear, experience in other cities indicated a subsidy range of between 74% and 82% of total costs (Shlay & Giloth, 1987).

Thus, once again, political calculations forced a shift in appeals and allegiances. As with the black community though, Byrne's initial promises enhanced a mobilization process that was well under way. Her about-face merely fanned the fires of activism and opposition. Not only did Byrne lose the election, but also her World's Fair went down to a stinging defeat at the hands of the very neighborhood movement she initially encouraged. The size and projected cost of the Fair shifted the terms of the downtown-neighborhoods debate, giving the latter groups the upper hand.[6]

Unfortunately for Byrne, the irony continued: Her alliance with the machine was not as productive as hoped. In its weakened and internally divided state, it could not provide the centralized control necessary to execute large-scale development projects. And it did not prevent her worst fear—a primary challenge from Richard M. Daley.

THE WASHINGTON REVOLUTION:
REFORM POLITICS AND PROGRESSIVE POLICY

If 1972 was a watershed year in Chicago politics, 1983 was the year that the dam burst wide open. The "city not ready for reform" elected a reform mayor; the city, fueled by racial concerns, elected its first black mayor. And this was just the beginning.

Harold Washington carried the electoral revolution into office, declaring all-out war on some long-standing practices. Specifically, the administration sought (a) to broaden the decision-making process beyond the "selective insiders," (b) to replace the machine's practice of dispensing individual benefits with a collective orientation, and (c) to substitute balanced development for the progrowth regime's corporate center strategy. Thus, the administration sought a marriage between reform politics and progressive policy.

In many ways, the administration was a culmination of key political developments in Chicago, particularly the machine's decline and the ascendancy of new organizational forces. Washington's departure from his predecessors was an acknowledgment of these changes and an attempt to forge a new order. It was also a response to the resource problem: The discretionary resources that fueled machine politics and progrowth policies had sharply contracted.

At the most fundamental level, Washington sought to merge the electoral and policy arenas, a feature evident in his campaign. His use of community-based organizations rather than the ward organizations was a signal that policy and positions were replacing patronage and precinct captains as major support-getters.

In contrast to his predecessors, Washington turned his electoral coalition into a governing coalition; rather than co-opt elites, he incorporated groups. Unlike Daley, Washington lacked the "selective incentives" required for such co-optation. A combination of antipatronage rulings, the extension of collective bargaining rights to the majority of city workers, budgetary shortfalls, and Washington's reform credentials bound him structurally and symbolically. Unlike Byrne, who did begin on a reform course only to retreat early in her administration, Washington had a committed bloc of voters in his black base. This support gave Washington a measure of electoral security that Byrne lacked.

Turning the electoral coalition into a governing coalition shifted the accommodation process from one of individual benefits to one of collective, or policy, benefits. The incorporation of neighborhood groups into the decision-making process was further reinforcement. The administration's first major policy document, "Chicago Works Together," was the result of numerous community forums and public hearings that relied on the extensive network of neighborhood organizations. Issued in 1984, it emphasized neighborhood participation, balanced development, affirmative action, open and fair processes, job creation, and redistributive policies (City of Chicago, 1984).

The breadth of this document and the administration's emphasis on collective benefits necessitated institutional change. Chicago's caretaker government had to be transformed into a policy-making body. Developing this capacity required the reform of existing institutions and the creation of new structures. The first was necessary to achieve equity in distribution, as opposed to the more arbitrary allocation patterns of the machine. New structures were necessitated by new economic realities that often outstripped the capacity of Chicago's patronage-bloated bureaucracies.

While Washington suffered many defeats, he made major inroads in the areas of education and economic development. These two areas probably best highlight the ideological shifts and the accompanying political, policy, and institutional changes of the Washington administration.

A major priority of the Washington administration was education. Although passed after his death in November of 1987, the educational-reform bill was a product of his administration's policy and lobbying efforts.

Heralded as the most innovative educational reform in the country, the measure featured a decentralization of decision-making powers to the communities through locally elected school councils. Underlying the entire measure was the changed view of government: Education is not a routine service to be neglected in an oversized bureaucracy, but rather a key element in the city's vitality. Procedurally, the changes were equally radical; centralized administration would be pushed aside for decentralized governance based on citizen participation. Hence, the electoral and policy politics converged.

Washington's initiatives in the area of economic development were equally bold in their challenge to long-standing practices. Economic realities, combined with the new ideology, propelled radical policy departures. The result was a reconfiguration of political and institutional practices.

Beginning in the 1970s and escalating in the 1980s, Chicago's economy took a nosedive. By 1982 the number of manufacturing jobs in Chicago was at its lowest level since the nineteenth century. Even the suburbs experienced losses in this category, reversing a 30-year growth trend (Weiss & Metzger, 1989). Alongside the job loss was population decline; the city lost more than 600,000 people between 1950 and 1980. Thus, the assumptions underlying the progrowth strategy—continued employment and population growth—no longer applied. The new economic problems required a broader view of economic development, a factor not lost on the city's movers and shakers. In a 1984 report issued by the Commercial Club (a major civic organization), strong calls were made for job creation and business promotion strategies (Commercial Club of Chicago, 1984).[7]

While Washington was hampered in the first three years of his administration by a belligerent City Council led by leaders of the old machine faction, he did secure innovations in the planning process. Using constituent pressure, a ward remap, and a special Council election that gave him a legislative majority, Washington altered certain practices in very fundamental ways.

The most significant change was the inclusion of neighborhoods into the economic development equation and women and minorities into the contracting process. Through the redirection of CDBG funds to neighborhoods, the passage of a major bond issue for neighborhood improvements, the negotiation of linked development agreements, and the incorporation of replacement housing provisions for projects that cause displacement, Washington redefined economic development in Chicago. Through the adoption of set-aside programs for minority and female-owned firms on city contracts, the administration broke the insiders' monopoly that had resulted in a highly inequitable system.

Under the machine-supported progrowth regime, the neighborhood experience ranged from public neglect and private disinvestment to urban renewal and substantial dislocation. For some neighborhoods, the Washington administration's bond issue resulted in the first capital improvements since the New Deal. Thus, in broadening the policy arena, the administration shifted the role of city government from supporter of economic elites to protector of the disenfranchised.

Although many business leaders recognized the need for policy change, they were not prepared for a new set of rules. Chicago's business community had been weaned on machine politics, closed-door processes, and insider privileges. Washington's policy platform threatened all of that and more. The calls for balance between neighborhood and downtown threatened the hegemonic control over economic development to which downtown business was accustomed. It also meant a shift in public investment from downtown to the neighborhoods. Citizen participation and neighborhood planning were direct attacks upon the ward organizations. Open and fair government threatened the very lucrative contract process, while affirmative action was yet another assault on that beleaguered Chicago industry—patronage.

That Washington faced serious political battles is thus not surprising. His successes, however, become all the more important in that they introduced the concept of balance into a highly uneven economic-development process and inclusion into a closed policy-making process. With government as the promoter of these changes, Washington redefined the role of government from caretaker and supporter of private economic growth to activist body with responsibility for shaping overall policy. Moreover, he moved it away from the informal, arbitrary dispenser of favors that it had been under the machine to a more formalized, professionalized, and nonpartisan entity. Thus, what began as electoral appeals in the Byrne campaign became policy benefits in the Washington administration. This "institutionalization" of benefits shifted political discourse.

RACE VS. REFORM VS. MACHINE: CHICAGO POLITICS, 1987 AND BEYOND

Washington's untimely death in November of 1987 left much of his agenda unfinished. It also left a political system up for grabs. In this sense, it resembled the aftermath of Daley's death 11 years earlier. The context, however, was markedly different. What Harold Washington and time had opened up would not be closed. Policy issues are a regular part of public debate; community-based organizations are important and respected actors; and "reform" has worked its way into the vocabularies of most people.

These significant developments, however, compete with a stronger force—race. Race has replaced reform as the unifying issue in elections. After Washington's death, white aldermen, realizing they lacked the votes to elect "one of their own," allied with several black aldermen to elect Eugene Sawyer over Timothy Evans as mayor. Although both are black, Sawyer was perceived as weak, manipulatable, and most important, beatable in a citywide election. Evans, on the other hand, was depicted as the heir apparent to Washington. Selecting Sawyer over Evans split the black community. When the special election was held in 1989, it was the black version of 1983: Two black candidates split the black vote, paving the way for a white victory— Richard M. Daley won the Democratic primary against Sawyer and went on to beat Evans, who ran on the newly formed Harold Washington Party, in the general election.

Although the campaign lacked the level of racial rhetoric that characterized Washington's two elections, particularly in 1983, it was clearly a black-white election. Neither Sawyer nor Evans came anywhere near the level of white support, or even black support, that Harold Washington received. The divisions and demobilization within the black community, combined with the larger racial divide, damaged reform's electoral chances. The white lakefront liberals, along with some parts of the Hispanic community, which were in the Washington coalition, jumped on the Daley bandwagon.

A NEW DALEY?

While reform languishes in the background, the city has a mayor trained in the Chicago school of machine politics. Moreover, many of his key financial backers are alums of the same school. Given the high cost of media campaigns, this is not an insignificant factor. On the other hand, many of Daley's voting constituents expect a continuation of the policy benefits they received during the Washington administration. Given the composition of his coalition, it is not surprising that Daley's actions reveal fundamental tensions between machine and reform approaches to governance and between pro-growth and progressive approaches to policy.

Following his father's advice that "good government is good politics and good politics is good government," Daley has continued with many of Washington's progressive initiatives, securing passage in the City Council of measures on which Washington was defeated. Among Daley's legislative successes is a series of measures and ordinances to assist community-based organizations in providing low- and moderate-income housing. Daley also assembled a new minority set-aside program that would comply with the Supreme Court's standards in the 1989 *Richmond* case (*Richmond v. Croson,*

1989). Several of his appointments, particularly housing and public housing, actively promote progressive policies, not the least of which is tenant empowerment in public housing.

On a procedural level, Daley has made strong reform overtures. The administration conducted public meetings on its capital improvements budget. It also has adopted Park Advisory Councils to serve as mechanisms for public input. The latter is especially significant since the Park District under Richard J. Daley was notorious as a patronage dumping ground, and it was the first major department on which Harold Washington unleashed his reforms.

Alongside these progressive reform measures stand strong reminders of yesterday's policies and approaches. Daley's proposal for a third airport in the city's Southeast Side is urban renewal with a vengeance. Daley's insistence on that site, despite neighborhood and economic soundness objections, is a throwback to the machine style of centralized decision making.[8] It also mirrors the practice of economic development at any cost. If constructed, the airport will force the relocation of approximately 8,500 families and 9,000 jobs (Ferman & Grimshaw, 1991).

While many of Daley's policies reflect a dichotomous world view of neighborhoods versus downtown, the realities of Chicago's problems and political landscape are quite different. The fallout from economic restructuring, the decline of federal funds, and the increase in local fiscal stress have created new problems, as well as new political alliances. Community-based organizations, in alliance with corporations, have taken the lead in affordable-housing initiatives, neighborhood reinvestment strategies, and the implementation of local school reform. Community-based organizations, often in conjunction with religious institutions, are launching major assaults on neighborhood drug and crime problems.

Chicago is at a crossroads. The current administration is straddling the convergence sought by Washington and the divergence maintained by Richard J. Daley. The administration is much more policy oriented than the latter, but in contrast to the former, it is an orientation that divides the city between downtown and the neighborhoods. Thus, the orientation resembles that of the progrowth regime with policy substituting for patronage as a buffer for economic elites.

In managing these tensions, Daley has benefited from the demobilization within the black community. Black mobilization was the central factor in undermining the old progrowth regime. The neighborhood movement is prone to divisions along racial, ethnic, class, and territorial lines, most of which the Daley administration has exploited. In the absence of a loyal opposition and with a near monopoly on big campaign contributions, Daley can restore the growth coalition and weather the rumblings of a divided neighborhood sector. Thus, as in the past, the relationship between the electoral and policy arenas will influence the direction of Chicago politics and policy in the 1990s.

NOTES

1. These books include: J. Allswang, *A House For All Peoples*, (1971); A. Gottfried, *Boss Cermak of Chicago*, (1962); T. Guterbock, *Machine Politics in Transition*, (1980); M. Meyerson and E. Banfield, *Politics, Planning and the Public Interest*, (1955); J. Tarr, *Study in Boss Politics: William Lorimer of Chicago*, (1971): L. Wendt and H. Kogan, *Bosses in Lusty Chicago*, (1967).

2. Books on Richard J. Daley represent a cross section of scholarly materials and journalistic accounts. They include: E. Banfield, *Political Influence*, (1961); E. Kennedy, *Himself*, (1978); L. O'Connor, *Clout*, (1975) and *Requiem*, (1977); M. Rakove, *Don't Make No Waves, Don't Back No Losers*, (1975); and M. Royko, *Boss*, (1971).

3. Daley did approve collective bargaining for the teachers' union, but it was the only such agreement.

4. The court ruled in 1970, but it was not until 1972 that the city signed a consent decree.

5. For excellent analyses of the role of race in Chicago politics, see William J. Grimshaw, "The Daley Legacy: A Declining Politics of Party, Race and Public Unions," in Gove and Masotti, Eds. *After Daley: Chicago Politics in Transition*, 1982, and Paul Kleppner, *Chicago Divided: The Making of a Black Mayor*, 1985.

6. For an excellent detailed analysis of the challenge to the Fair, see Robert McClory, *The Fall of the Fair: Communities Struggle for Fairness*, 1986 (The Chicago 1992 Committee).

7. The Commercial Club is one of the oldest civic organizations in Chicago. It sponsored Burnham's plan for the city of Chicago. Proposed in 1909, it was the first comprehensive plan for city development in the nation.

8. If the third airport destroys Midway Airport, as some have argued it will, the Southwest Side of the city (where Midway is located) will experience serious economic dislocation.

REFERENCES

Allswang, J. (1971). *A house for all peoples*. Lexington: University of Kentucky.

Banfield, E. (1961). *Political influence*. New York: Free Press.

City of Chicago. (1984). *Chicago works together: 1984 development plan*. Chicago: Department of Economic Development.

Commercial Club of Chicago. (1984). *Make no little plans: Jobs for metropolitan Chicago*. Chicago: Commercial Club.

Ferman, B., & Grimshaw, W. (in press). Housing Chicago style: A tale of two cities. In K. Wong (Ed.), *Politics of policy innovation in metropolitan Chicago*). Greenwich, CT: JAI.

Gosnell, H. (1937). *Machine politics: Chicago model*. Chicago: University of Chicago Press.

Gottfried, A. (1962). *Boss Cermak of Chicago*. Seattle: University of Washington.

Green, P. (1987). Michael Bilandic: The last of the machine regulars. In P. Green & M. Holli (Eds.), *The mayors: The Chicago political tradition* (pp. 164-171). Carbondale, IL: Southern Illinois University Press.

Greer, J. (1986). *Capital investment in Chicago: Fragmented processes, unequal outcomes*. Chicago, IL: The University of Chicago, Center for Urban Research and Policy Studies.

Grimshaw, W. J. (1982). The Daley legacy: A declining politics of party, race and public unions. In S. Gove & L. Masotti (Eds.), *After Daley: Chicago politics in transition* (pp. 57-87). Urbana, IL: University of Illinois Press.

Guterbock, T. (1980). *Machine politics in transition*. Chicago: University of Chicago

Holli, M. (1987). Jane Byrne: To think the unthinkable and do the undoable. In P. Green & M. Holli (Eds.), *The mayors: The Chicago political tradition* (pp. 172-182). Carbondale, IL: Southern Illinois University Press.

Kennedy, E. (1978). *Himself*. New York: Viking.

Kleppner, P. (1985). *Chicago divided: The making of a black mayor*. Dekalb, IL: Northern Illinois University Press.

Lowi, T. (1968). Gosnell's Chicago revisited via Lindsay's New York. In H. Gosnell (Ed.), *Machine politics: Chicago model* (2nd ed.) (pp. v-xviii). Chicago: University of Chicago Press.

McCarron, J. (1990, April 1). A sneak peek at what 1990 Census will tell us. *Chicago Tribune*, pp. 1, 9.

McClory, R. (1986). *The fall of the fair: Communities struggle for fairness*. Chicago: The Chicago 1992 Committee.

Meyerson, M., & Banfield, E. (1955). *Politics, planning and the public interest*. New York: Free Press.

O'Connor, L. (1975). *Clout*. Chicago: Regnery.

O'Connor, L. (1977). *Requiem*. Chicago: Contemporary Books.

Rakove, M. (1975). *Don't make no waves, don't back no losers*. Bloomington: University of Indiana.

Richmond, VA v. G. A. Croson Co., 109 S. Ct. 706(1989).

Royko, M. (1971). *Boss*. New York: Signet.

Shlay, A., & Giloth, R. (1987). The social organization of a land-based elite: The case of the failed Chicago 1992 world's fair. *Journal of Urban Affairs 9*, 305-324.

Stone, C. (1989). *Regime politics: Governing Atlanta, 1946-1988*. Lawrence, KS: University of Kansas Press.

Squires, G., Bennett, L., McCourt, K., & Nyden, P. (1987). *Chicago: Race, class and the response to urban decline*. Philadelphia: Temple University Press.

Tarr, J. (1971). *Study in boss politics: William Lorimer of Chicago*. Urbana: University of Illinois.

Weiss, M., & Metzger, J. (1989). Planning for Chicago: The changing politics of metropolitan growth and neighborhood development. In R. Beauregard (Ed.), *Atop the urban hierarchy* (pp. 123-151). Totowa, NJ: Rowman and Littlefield.

Wendt, L., & Kogan, H. (1967). *Bosses in lusty Chicago*. Chicago: Bobbs-Merrill.

5

Detroit:
From Motor City to Service Hub

WILBUR RICH

THE POLITICAL HISTORY of Detroit reflects its multiethnic heritage and evolution as a center of industrial unionism in America. Unlike Carl Sandburg's Chicago, a city with big shoulders, Detroit has always had small shoulders on which ethnic groups found it hard to co-exist and were doomed to spatial conflict. In 1925 and 1943 two major race riots were caused, in part, by housing competition, unemployment, and poor police-community relations. After World War II the city politicians spent considerable time trying to keep ethnic whites and blacks apart. Ethnic groups increasingly saw city politics in zero-sum terms.

In *Big City Politics* (Banfield, 1965), Detroit politics was characterized as a balancing act between the array of interest and ethnic groups. A 1966 survey found ethnic whites were keeping their cultural identities and religious affiliations (Laumann, 1969). Agocs concluded that the "melting pot metaphor is not descriptive of Detroit's population" (Agocs, 1974, p. 404). The forces of assimilation did not take hold, but the toll of separateness and inequality did affect the life chances of the growing black population.

DETROIT'S CAMELOT

The election of Jerome Cavanagh as mayor in 1962 marked a critical shift in Detroit's political history. It came at a time when the United Automobile Workers (UAW) was trying desperately to consolidate its relations with City Hall. The union backed the incumbent, Louis Miriani, because he looked like a sure winner. The business community backed Miriani for much the same reason. Cavanagh, however, was able to recruit black leaders to his campaign by promising them a role in the new administration. The unexpected bolt by black labor activists from their union's official endorsement changed the equation of the mayoral race. Despite all the energy and resources of the labor leadership and strong support of business marshaled behind Miriani, Cavanagh won. The victory inflated the reputation of Cavanagh (Jerry the Giant Killer) and forced the unions to make their peace with this young upstart (33 years old).

Detroit, like the rest of the nation, was caught up with the Camelot analogy of the Kennedy years. The analogy was not lost on Jerry Cavanagh, and he sought to identify with the Kennedy mystique. Conot observed that "there was the illusion that Jerry and Mary were a replica of John and Jackie" (Conot, 1974, p. 493). Conot concluded that Cavanagh had "charm enough to reduce a cobra to a lap pet and a wit sharp enough to hone a rusty blade for a television commercial. He was living the American success story: overalls to pinstripes in one generation" (Conot, 1974, pp. 447-448).

Cavanagh nurtured the notion that he was somehow responsible for the economic resurgence of the city. Indeed, he personally intervened in the fight to repeal the state tax on locally assessed personal property—such as tools, dies, and fixtures—in exchange for a new Chrysler plant in the city. There was also a downtown building boom, and the mayor received credit in the press for changing the "embattled city to the city of promise" (Conot, 1974, p. 491).

The governing regime under Cavanagh remained the same as it had under Miriani. Lynn Townsend at Chrysler, Henry Ford II at Ford, and James Roche at General Motors, the corporate giants of their time, had what Clarence Stone later called "preemptive powers," that is, the capacity to occupy, hold, and make use of strategic positions in the decision-making process (Stone, 1988, pp. 91-92). Because of who they were and what they represented, they could exert influence over any planning meeting without being there physically. City politicians prided themselves in making The Big Three's work in the city pleasant. The young man who came to the mayor's office as a dashing reformer had become a team player.

The riots ended the career of Jerome Cavanagh and Camelot but did not diminish the influence of the governing regime. They simply changed the tactics of the regime. Its members could no longer keep a low profile in city politics. They created and funded New Detroit Committee and Detroit Renaissance as their principal vehicles for socioeconomic recovery of the city and as a more direct way of interacting with black leaders (Rich, 1989). Black politicians and professionals simply supplanted white liberals as intermediaries between the business elite and the black masses, without any fundamental changes in the governing of the city.

In the aftermath of the riots, tensions abounded in the city as whites fled to suburbs and blacks continued to demand full participation in the city's politics and economy. Detroit's political environment had become completely unbalanced. The Irish and Polish politicians lost their power base because their fellow ethnics dispersed into the suburbs. The rampant flight of ethnic groups left a minority of liberals, small pockets of Southern immigrants, and elderly residents as the remaining white population. The 1970 Census found that the black population had risen to 43% of the city's population.

Although the black candidate for mayor, Richard Austin, had lost by 1% of the vote in 1969, the trends pointed toward a victory in 1973. In 1973 the first black mayor, Coleman Young, was elected as a biracial coalition candidate, and the process Eisinger (1980, p. 5) called "ethnoracial transition" began. There are serious problems, however, with the composition and stability of his governing regime. How has the Coleman Young era enhanced or hindered the economic development of the city? Can the proposed 1990 redevelopment projects facilitate the city's service economy ambitions? Is the Detroit political regime an effective one? This chapter will examine these economic and political questions about Detroit's governing regime since 1960.

THE DECLINE OF AUTOMOBILE MAKERS

Detroit was once a center of American industrial power, with automobile jobs drawing people from all over the world. The plight of automobile sales still drives the Michigan economy. When cars are selling well, all is well in Michigan and vice versa. Detroit is at the mercy of consumer preferences, gas prices, and the health of the economy.

On the whole, however, the automobile industry has lost its engineering edge to foreign automobile makers, and the manufacturers have managed to alienate the consumer. American consumers believe that foreign cars are better made. In 1965, the Big Four built 9,332,000 cars and 1,787,000 trucks. Now American Motors has merged with Chrysler, and in 1988 only 7,116,000 cars and 4,080,000 trucks were built, a loss of 2,000,000 cars. The 4,000,000 trucks represent the peak of truck building by the industry, and it is doubtful that the industry will ever match its 9,662,000 cars of 1973. The streets of Detroit are full of Japanese- and German-made cars. Mercedes Benz is now the preferred luxury car in the city, with drug dealers, professionals, and suburbanites keeping the metro Mercedes Benz dealerships in business.

The continuing loss of market share has given a rude awakening to the leaders of the Big Three. The corporate leaders have become preoccupied with steering their companies out of the present dilemma. The automakers are making adjustments in their investments. Chrysler has bought American Motors; created a new nameplate, Eagle; and is looking for a foreign partner. Despite the salesmanship of its chairman, it is still losing money. General Motors has bought Electronic Data Systems, a computer company. Ford Motors, once the most profitable of the three, has invested in the aerospace industry. Meanwhile, the financial world, which turned upside down with the Wall Street crash of 1987, was very worried about the price of gas during the Persian Gulf crisis. The automobile industry finds itself adrift, surrounded by dubious consumers in an aggressive international market. Detroit finds itself an aging city with a central industry in peril and a core population of essentially unskilled workers.

SUBURBANIZATION OF DETROIT

In 1954 the first shopping mall (Northland) was built in a small town called Southfield. No one would have imagined that 30 years later the mall would be the primary retailer for blacks in Detroit or that it would evolve into a hangout for black teenagers. Today metro Detroit boasts several shopping malls, the largest of which is located in Dearborn. The Fairlane Shopping Center, with the help of Interstate 94 and the Southfield Freeway, is accessible for most of Detroit's residents. With the exception of the New Center area, there are few places to buy department-store goods in Detroit. These shifts in shopping patterns underline the fundamental demographic changes in the tricounty area. New, smaller cities have grown up at the expense of Detroit and act as drains for Detroit's population.

Between 1970 and 1980, Detroit lost 394,147 white residents. Overall the city lost 20.5% of its population by 1980. In the 1990 Census the Detroit population sank to barely 1 million, the lowest figure since 1920. Between 1960 and 1970 Troy grew by 103.2% and added another 70.2% in 1980. Southfield grew by 119.9% between 1960 and 1970. Southfield ceased to grow, however, after 1970 as it gained a reputation for welcoming blacks. Recently, racial tensions have flared at Southfield High School, which is now 50% black, and the mayor of Southfield has had to apologize for expressing reservations about Southfield becoming a more integrated city. Meanwhile, many white professional firms have moved farther away from the Detroit city line and nearer to more affluent suburban areas. Little Novi now has grown into a complex of hotels and office buildings and grew by 133% between 1970 and 1980.

In the 1980s Southfield lost its progrowth reputation, first to Troy and then to Auburn Hills. Although a small community of 16,000, Auburn Hills increased its population by 10% between 1980 and 1990. As the new home of the Detroit Pistons basketball team, its plans to build a 200-store discount megamall at the junction of the new Interstate 694 would put all metro retailers at risk. The proposed mall will be able to undersell most, if not all, existing malls and will provide an incredibly large parking area.

The result of this outmigration practically has eliminated the ethnic mix of Detroit itself. The city is now 76% black. The lower-class Polish and Irish enclaves are gone, and Jews have moved to Oak Park and other suburbs. The remaining whites tend to be the elderly, white liberal riverfronters and integrators (a term used to designate whites who prefer living in integrated neighborhoods, e.g., the STAYERS Club in Rosedale Park), and white enclaves around the University of Detroit.

Native Detroiters usually start their conversations with "I remember when. . . ." They complete the sentences with stories of abundant work on the assembly lines. Theirs is a Detroit of humming assembly lines, wide-body sedans, jobs going begging, and single-family homes. This Detroit is gone. Gone is J. L. Hudson's Department Store, once the largest department store in Michigan. The Kern Block, once the site for Kern's Department Store and thought for years to be a candidate for a historic district and boarded for nearly a decade, has finally been dismantled. The old Greyhound station has moved farther away from downtown.

Stores on Woodward Avenue, the city's main street, cannot make a profit for the lack of traffic. The downtown merchants cannot conduct business next to abandoned spaces. The People Mover, designed to attract more traffic to the central business district, has not been able to pay for itself, with only tourists, deliverypersons, and office workers as riders. Just as the People Mover has changed the physical appearance of downtown, the retail sales function is also slowly changing. The city administration has invested heavily in the entertainment and convention business. It also competes for office space. Recently, the city has supported the erection of two large office buildings, the first since the completion of the Renaissance Center, but these new buildings are being built in the face of a low-market rental rate and a glut of office space in the suburbs.

Detroit's redevelopment strategy of the 1980s has been only partially successful. The city continues to lose office space to the suburban communities. Such suburban cities as Southfield and Troy are enjoying a boom in office building. The Cobo Hall convention center has been expanded, and new hotels have changed the skyline of the city, yet there is still a lack of hotel rooms for conventioneers.

The shrinkage of manufacturing industries has undermined the structure of the city workforce. No longer can unskilled laborers look forward to a job on the assembly line. Many of the new blue-collar service jobs offer the salaries and benefits of the automobile industry, but the service economy has not been a boon for black inner-city workers. The new jobs are in the suburbs, adding to the transportation cost for workers.

Although economic statistics are kept for only the entire metro Detroit area, they tell a lot about the changing nature of the workplace. Since Young's election in 1973, employment in the trade sector has increased from 323,000 to 447,000; real estate jobs from 75,000 to 110,000; and service employment from 273,000 to 480,000. In other words, the nonmanufacturing sector of the economy has outgrown the manufacturing sector. At the time of Banfield's book (1965), manufacturing jobs accounted for 42.4% of the area's jobs. In 1988 it represented only 24.2% of the job market. The effect has been devastating on the black and poor populations, which continue to have the highest unemployment figures in the region. In 1988 unemployment was 7.7% for the region and 15% for Detroit, an increase of 4.5% for the region since the 3.2% of 1965.

Yet the combined aggregation of the city and the suburbs keeps the hourly wage averages for the city among the highest in the nation. The 1965 average hourly wage was $3.39, as compared with a national average of $2.61. The 1988 average hourly wage was $14.44, as compared with a national average of $10.17. The average hourly wages are still inflated by automobile workers' salaries.

The city ended the 1980s with some hopeful signs. It built a People Mover, despite the Reagan administration's lack of interest in urban transportation policy. Stroh Brewery Company made a major commitment to revitalize idle riverfront properties. Mike Illich, the Little Caesar's pizza mogul, renovated the Fox Theater and has nearly completed his plans for locating his head-quarters in downtown Detroit. Chene Park is carving its place among amusement parks. In addition, local entrepreneurs have built riverfront apartments and condominiums.

Yet, there is growing dissent over the administration's redevelopment policy. The erection of building cranes has not impressed the business community in metro Detroit. In a recent poll of area businesspersons conducted by the Greater Detroit Chamber of Commerce, more than 80% expressed reservations about the business climate in the city (Deck, 1990). Only 15% thought the climate was getting better. Although the crescent around the city is enjoying a boom in suburban hotel and motel space, the image of downtown is of irreversible decline.

In 1968 the Kerner Commission predicted that there would be fewer tax dollars available as the middle class and businesses move out. The decline in tax revenue will come at a time when more services will be needed to serve the increasing low-income residents. An inflation problem and growing dissatisfaction with the performance of cities will occur.

The once-invisible poor of the 1950s and 1960s are seen now in most Detroit neighborhoods. Their welfare checks and food stamps have become the currency of large sections of the city. Coupled with the underground economy of the drug trade, with its flashy cars and expensive jewelry, Detroit neighborhoods have become a study in contrasts. These individuals cannot support a downtown retail market. The downtown area, once the center of southeastern Michigan, has become a haven for street people and the homeless.

ECONOMIC DEVELOPMENT PLANS

Politics in Detroit is a particularly fascinating example of the cooperation between black political leaders and white business leaders. The evolution of this regime from a shaky beginning into a complex mutual-benefit relationship is proof that black politicians can play a major leadership role in the economic development of American cities. The new coalition includes members of the old regime and the new service moguls. Men like Max Fisher, investor and financier; Al Taubman, shopping-mall developer; and Peter Stroh, chairman of Stroh Brewery, along with newcomers Tom Monaghan, owner of the Domino's Pizza empire; Mike Illich, owner of Little Caesar's and the Detroit Red Wings hockey team; Frank D. Stella, of Stella Products Co.; and Frank Hennessey, of the Handleman Company, have joined with bank presidents and heads of the local utilities to form the business elite.

The corporation elites, who consist of Robert Stempel of General Motors, Lee Iacocca of Chrysler Corporation, Harold A. Poling of Ford Motor Company, Robert Dewar of K-Mart Corporation, and James Cordes of ANR Pipeline Company, rarely take an active role in city politics, although this has changed somewhat since Mayor Young invited General Motors to build the controversial Poletown Plant (Jones & Bachelor, 1986). Swanstrom (1985) has divided downtown businesspersons into mobile and immobile elites. The immobile interest is invested primarily inside the metro Detroit area, whereas the mobile elites represent multinational enterprises. The corporate elite of Detroit consider themselves world players, with their headquarters in Detroit but much of their business operations outside the city. Such local elites as banking, utilities, and real estate interests that are closely tied to the city tend to be more intensely involved in local development.

Mayor Young entered the 1990s with two controversial development plans on the drawing board. The first of these, the expansion of the City Airport, located on the East Side near downtown, will be a $650 million project covering 1,250 acres, a new 7,400-foot runway, and a new terminal. The plan calls for an expansion to four times the present size by 1998. The new expansion could accommodate 100 flights a day and several commercial airlines. The city has promoted the plan as an economic development strategy for the city's eastside.

The project is bigger than the Poletown GM plant and the Chrysler-Jefferson Plant. To expand the airport, 3,600 occupied dwellings will have to be evacuated, another 60 vacant homes will be demolished, and over 50 small businesses will have to be relocated. Residents are concerned about the increase in traffic, noise, and air pollution. The expansion of the airport will not cause competition with Metro-Airport, the county-owned facility, but will provide a faster service for those who live on the East Side of metro Detroit and downtown. The first expansion of the runway was met by protest over the removal of a cemetery. The announcement of a larger expansion brought out the mayors and residents of East Side suburbs in protest. Southwest Airlines has repeated its threat to move if it doesn't receive a longer runway and larger terminal.

The second controversial project is the proposed demolition of Ford Auditorium for a new headquarters for Comerica Bank. This is projected to be a 30-story building located on the city's riverfront. Comerica initially agreed to buy the auditorium for $7.5 million and raze it. They also agreed to lease the existing underground parking lot for $76,750 a year. Comerica would receive an $18-million no-interest development loan from the city, with payment to be made for 28 years. Hines, Inc., the developer, and the bank stated that this arrangement was necessary in order to attract the private investor for the needed $150 million. They cited other cities that have subsidized office buildings and argued that Detroit as a low-rental-rate market is in competition with these cities for office space.

This deal caused a firestorm of protest. A group of civic leaders formed the Ad Hoc Coalition for Fair Banking Practice in Detroit. Headed by Arthur Johnson, president of the Detroit NAACP, the group met with Comerica and persuaded the bank to pay 4.75% interest on the city loan and to pay the city $1.5 million as down payment on the site. The bank decided to buy the parking garage for $724,000. The $5 million debt of the garage would be assumed by the Downtown Development Authority.

This so-called sweetening still did not sit well with the public. Citizens protested the demolition of the Ford Auditorium, a $2.5 million gift from the Ford family. William Clay Ford publicly protested the planned project and vowed never to make another gift to the city. He was joined by supporters who wrote letters to the editor.

The Auditorium is a very sentimental building in Detroit's history as the site for many events, particularly high school graduations. It never broke even, however, and was heavily subsidized by the city. The struggle and debate over the project was the most heated in years. The bank and the developer heavily lobbied the city for an affirmative vote and threatened to move to its lot in Auburn Hills, a new growth area in metro Detroit, if the project was rejected. This relocation would mean the loss of 1,100 headquarters jobs from the downtown area. Restaurants and shops would lose even more traffic. Facing that prospect, the City Council approved the project by a vote of 6 to 3. The fight did not end there, however, as the three councilors voting no threatened to join a lawsuit with retiring Congressman George Crockett.

The continuing Comerica fight highlights the problems of cities like Detroit. When the same developers made a deal to build a large office building, One Detroit Center, behind the city-county building and close to the People Mover, there was little protest even though demolition of the old Greyhound station and a motel were involved. The location is so choice that the developer invested $50 million of its own money to build it.

The increased interest in downtown locations has transformed the mayor into the chief city salesman and agenda organizer. Coleman Young has been so personally involved with these projects that he is thought to be the leader of the governing regime, not just a mere partner in a social-production enterprise.

WHITHER THE MAYOR-CENTERED COALITION?

Framing the agenda is what the mayoralty is all about. Within City Hall, economic as well as other priorities are established. In this task, mayors have had few rivals. The City Council, lacking a professionalized staff and having a weak political base (at-large constituencies), is not in the position to challenge the mayor's policy initiatives.

Young has been able to get things done despite being unpopular among reporters. Part of the reason business as usual has continued under Young is that the loose coalition of reformers, corporate leaders, and new black leadership came apart after his election. Young has encouraged the business elite to deal with him directly. Accordingly, Young has become the leader of the black participants in the coalition, while still retaining his powerful role as mayor. Young does not have the same type of charm as Cavanagh, but he is less ambitious than his famous predecessor. The economic leaders discovered that "Coleman Young had one agenda: redevelopment of Detroit" (Rich, 1989, p. 139). Although the support of the economic leaders was necessary for Young's redevelopment project, it was not sufficient.

Since the Charter Revision of 1918, mayors of Detroit have presided over a divided government. Cavanagh did not have complete control over the police department during the riots. Roman Gribbs, Cavanagh's immediate successor, did not have the support of the Civil Service Commission when he attempted to meet the demands of affirmative action in municipal departments. Both mayors needed more legislative authority to initiate and fund economic redevelopment projects.

Both Mayors Cavanagh and Gribbs fought for a new city charter to give the office of mayor administrative control of city agencies. A Charter Commission, appointed by Gribbs, recommended that the Civil Service Commission be replaced with a Department of Personnel, the director of which would be appointed by the mayor and serve at his or her pleasure. For decades Detroit mayors had complained about the power and independence of the Civil Service Commission. In 1973 the voters approved a new charter that authorized the Department of Personnel.

The strong mayor-council form of government, lauded by Banfield, became stronger with the new charter. Coleman Young was the clear beneficiary. Most of the important city department heads now serve at the pleasure of the mayor. This new authority enabled the mayor to promote his citywide affirmative-action program. The new charter also authorized the appointment of a deputy mayor and gave the mayor more authority over the police department. Despite the resistance of police unions, he has promoted affirmative action, including more women in command positions and as officers on the beat.

The new charter is not the only source of mayoral influence. The mayor may gain political power through informal means. Young has a wide network of friends and associates placed in strategic positions in the private sector and in government, and many of them got their jobs through mayoral contacts. Although most city jobs are covered by civil service, association with City Hall can facilitate one's career. In addition, Young has built political IOUs through his ceremonial roles. As head of the city, he performs a variety of social roles. He can ration his appearances at social events and gatherings so as to inflate the significance of his attendance. Groups know that he will attract media and good turnouts for events.

The mayor also uses the media to publicize his policies. Young's activities and behavior are considered newsworthy. He can dominate the local news and shake up Lansing with a news conference. In addition, his role as titular head of the Democratic Party makes his endorsement valuable. Unspent funds from his campaigns for mayor may be used legally to support other candidates for office and for charitable purposes, creating more political capital. He also may use party contacts to exert pressure on other officeholders to support his policies and proposals. The mayor also has contacts in the Republican Party that he has used effectively during the Reagan/Bush period (Rich, 1989).

OTHER GOVERNMENT ACTORS

Detroit is located in Wayne County and accounts for 47% of the county's population. In 1981 Wayne County was the first Michigan county to adopt a charter, a reform measure consolidating several far-flung and autonomous county government agencies under the new office of county executive. The first county executive, William Lucas, a black politician who made a reputation as a law-and-order sheriff who promoted more road patrol for small towns in the county, was elected over Democratic candidates. Many thought that Lucas would be a countervailing force against Coleman Young. This never happened, but Lucas was successful in anchoring the new elective office in the county governing process. Under his leadership the county sold its money-losing Wayne County Hospital and gained control over the one-time autonomous Wayne County Road Commission. Because both of these agencies were heavily unionized, these changes alienated most of the state labor leaders from Lucas.

In 1985 Lucas was persuaded to become a Republican, and in 1986 he left office to run for governor. This failed attempt at statewide office ended his career in Michigan. Edward McNamara, a former mayor of Livonia, a white suburb, was elected county executive. McNamara's biggest policy decision has been to seize control of the county jail from the sheriff. He also has promoted the expansion of Metro-Airport. Otherwise, he has maintained a low profile and has not asserted himself as the White Hope against Coleman Young.

In 1988 Wayne County experienced its first major fiscal crisis as a charter government. The county's new system of indigent inpatient health care, started by the Lucas administration, continued to have enormous cost overruns. Under the Indigent Health Plan, the state pays for indigent inpatient care and the county reimburses the state for the cost of hospitalization. Despite the fact that this system covers only poor with homes, the system has been flooded with patients. The deficit caused by the payment rose to $200 million, making it difficult for the county to preach fiscal integrity to Detroit.

The role of the county executive, however, has added a new element to Detroit politics.

The creation of new development authorities has added yet another new element. Since his election in 1973, Coleman Young's tenure has been dominated by economic development issues. The 1974 oil embargo exposed the automakers' competition with imports and threw the city into a deep recession. The mayor engineered a brief rescue from the city's fiscal problems in 1975, but in 1981 Young had to increase income taxes and to borrow $115 million in short-term notes. The two crises resulted from the city's loss of its tax base (Rich, 1989). Retail businesses were leaving the central business district, many middle-income residents were selling their homes and moving outside the city, and office rentals were down. The mayor found himself struggling to keep businesses and jobs in the cities.

In 1975 state legislation created the Detroit Economic Growth Corporation (DEGC). The new public authority was designed to implement development plans for the city. In 1978 the Downtown Development Authority (DDA) was created to oversee the development of the downtown area. The mayor serves as chair of a nine-member DDA board that has the authority to fund projects and to issue revenue bonds. The City Council must review and approve the DDA development plans. In the legislation two development districts were created, one covering most of downtown (between the two main freeways and most of the riverfront), the other covering the area around the 36th District Court Building.

The revenue source of the agency is the tax increment financing. Any property taxes left over from the established tax base are captured by DDA for development. So far the biggest project attempted by DDA has been the Millender Center, a hotel/apartment-retail complex adjacent to the Renaissance Center. The sister agency to DDA is the Economic Development Corporation (EDC), which focuses on industrial development. EDC is involved very heavily in the proposed new Chrysler plant. It has a separate board and also is chaired by the mayor. The DEGC serves as staff for both EDC and DDA.

The Southeast Michigan Area Rapid Transit (SMART), formerly Southeast Michigan Transit Authority (SEMTA), is the transit authority for the region. The People Mover was originally a SEMTA project, but it was transferred to the city (Rich, 1989). The new law creating SMART allows Detroit to keep a separate transit system and gives the mayor more authority over the regional system, but SMART still purchases equipment for the Detroit Department of Transportation and makes all regional transportation plans.

CONFLICT AND CONSENSUS

Nothing has changed so dramatically since the Banfield review as the array of interest groups. AFL-CIO COPE (Committee on Political Education) is not so ubiquitous in city politics as it was. The UAW CAP and its labor colleagues still have black staff members who devote considerable time to elective politics, but fewer blacks are active automobile workers. The Trade Union Leadership Council (TULC) continues to exert some influence in local politics. Horace Sheffield, however, is no longer active with the group and has his own organization called Detroit Association of Black Organizations (DABO). The Group on Advanced Leadership (GOAL) organization, once led by Rev. Albert Cleage, is now defunct and its members are organized as the Black Slate. As the political arm of the Shrine of the Black Madonna (formerly the Central Congregational Church), the Black Slate remains the last of the functioning black nationalist groups in the city. The winner of the 1988 Congressional Democratic primary, Barbara Collins, a member of the Shrine, should bring the group new prestige.

The seat of black organized political power is the 1st and 13th Congressional Districts, but their leadership has undergone generational replacement. The late Buddy Battle, one of the mayor's first labor supporters and longtime leader of the 1st District, has been replaced by Mrs. Mildred Stallings. Frazier Kemper is chair of the 13th District. These individuals do not possess the power of their predecessors. They are not contemporaries of the mayor and owe more to him than he does to them.

Conflict among the interest groups is beginning to surface. Normal generational changes are not going smoothly. Disputes arise about who is closest to the mayor and who should be considered as his heir apparent. Meanwhile, the white members of the city council are in the mid-60s or older, and their replacements most likely will be black. No black businessperson has emerged as the clear voice of black entrepreneurs. Blacks have yet to pool their resources for large projects as their white counterparts have.

Dissatisfaction continues about the role of the civil rights organizations and their leaders. Once challenged only by labor-organizing counterparts, their organizational missions are now widely questioned. New Detroit, Inc., founded after the 1967 riot, is now heavily devoted to service projects, overlapping with the Urban League. The NAACP chapter remains the largest in the nation but has lost its appeal to the youth. The ascendancy of Arthur Johnson, its former executive director and close advisor to the mayor, to the presidency of the NAACP has helped the organization's image.

The community of clergy remains fragmented along denominational and affiliational lines. Endorsements of the ministerial councils are sought eagerly by candidates, but few ministers actually are able to sway their congregations. Their greatest success was the defeat of the casino-gambling proposal, when they were helped by the implicit value conflict of their congregations.

Since 1976 three referenda on casino gambling have failed to win voter approval. The defeat in 1988 may be the last gasp of the casino-gambling proposal. Although not initiated by the Young administration, its defeat was associated closely with the mayor. The campaign against the casinos was opposed by a coalition of the Chamber of Commerce, the business community, black ministers, and community groups. Many in the opposition groups were members of the governing regime in good standing. The underlying causes of the casino defeat were significant not as a referendum on the mayor's leadership, but rather as an example of value conflicts among working-class voters (Rich, 1990).

The propensity to challenge such economic proposals as the airport expansion and new office buildings has added a new dimension to the city's politics. The opposition, however, remains very fragmented. No identifiable antidevelopment spokesperson has emerged.

CITY COUNCIL AND ECONOMIC DEVELOPMENT

The much-maligned nine-member City Council seems to be making a comeback as Young approaches the end of his career. The at-large nonpartisan election system, adopted in 1918, makes each race a popularity contest. Detroit is the largest city in the country to use this system. Under the system, the largest vote-getter is elected president. Name recognition and endorsement play a critical role in election, and incumbents rarely are voted out. In the last election John Peeples became the first councilor to be defeated in 24 years.

In a 1988 special election, voters approved a ballot proposal reducing the number of votes required to override a mayoral veto of the budget from seven to six votes. As a result, the mayor must be more conciliatory on budget issues. Currently there are three antiadministration councilors. The mayor can count on six votes on other issues, but budget approval has become a matter of intense negotiation. Council members usually disagree with the mayor's attempt to reduce spending for neighborhood groups, the most organized nonunion group in city politics.

In the 1989 election two new members were elected. Gil Hill, a part-time actor of *Beverly Hills Cop* movie fame and a former police inspector, was elected on his first attempt at elective office. The Rev. Keith Butler, pastor of the 4,000-member Word of Faith Christian Center and a major player in the anticasino drive, also was elected and became the first black Republican on the Council. Maryanne Mahaffey, one of four white councilors, received the largest number of votes in the at-large system and became President of the City Council.

The Council has tried to assert itself. As a forum for opposition groups who opposed the Chrysler Jefferson Avenue Plant, expansion of City Airport, and the Comerica Bank Proposal, it has attracted more attention. Although the Council controls the City Planning Commission, it does not develop development plans, and the lack of staff resources has left the Council in a reactive role.

Councilor Mel Ravitz, who gained fame as a poverty worker in the 1960s, has emerged as the chief critic of mayoral actions. Rev. Butler has joined him, along with new Council President Mahaffey. The alliance of two white liberals and a black conservative is unique, but the group is too divided philosophically to be called a working coalition. The other six members of the Council have supported the mayor on the big-ticket projects, with occasional dissent on other issues.

Dissent comes mainly from city councilors and community organizers who want the mayor to do more for decaying neighborhoods. This neighborhood coalition, although loosely organized, has created problems and opportunities for the administration. The coalition leaders blame the mayor, not the business community, for the lack of decent housing and joblessness.

At first glance this debate about neighborhood versus downtown development sounds like the normal discourse of insiders and outsiders. A closer look reveals an incredible amount of anxiety on the part of both groups about job losses in the city. Although the mayor's coalition wins most battles, their victories are given more legitimacy by defeating the opposition. Without such opposition the city would not be able to create the bargaining positions it needs in negotiation with business leaders. As Stone and Sanders, (1987) have reminded us, politics matters. On occasion, it can force both sides to compromise their positions for the good of the community.

DETROIT POLITICS IN TRANSITION

The Coleman Young era continues to be confronted with a hemorrhaging economy, an abandoned central business district, and massive unemployment. His response to these challenges was to redirect the city's resources to rebuild downtown, which has become a collection of abandoned stores and buildings. To support downtown revitalization, the new mayor wholeheartedly supported the growing service economy. This downtown strategy, backed by the real estate industry, construction unions, banks, and remaining downtown merchants, was aimed at creating a new and more diverse city economy.

Although until the mid-1980s the impact of the "new" economic development had been felt most keenly by downtown residents and merchants and by a small number of investors, no section of the city has been insulated from the externalities associated with downtown redevelopment. City politics is dominated by project development politics, as one downtown project plan after another is put through the maze of the city's politics. The politics of downtown development has eclipsed the ethnic conflict identified by Banfield (1965). The differences between 1965 and 1990 are more than simple generational replacement of political activists or the advent of more media-oriented politics. It is a change of political form and substance.

These changes and controversies have promoted a decline in confidence in the city government. In a 1990 telephone interview poll, Bledsoe found considerable dissatisfaction with local government since 1954 and traced a precipitous drop in positive assessments, until today, "Detroit city government receives a support score rating of negative 63%" (Bledsoe, 1990, p. 6). Even among Detroit blacks, city government today has a support score of minus 22, 11 points worse than a similar score for 1971. Bledsoe reports that the downward trend started before Coleman Young's election, suggesting that the city government's "troubles go beyond any one individual" (Bledsoe, 1990, p. 26).

Coleman Young has dominated Detroit politics for over 16 years. He has done so with a mayor-centered coalition. With the help of public authorities and his own personal connections with investors and developers, he has been able to determine the political and economic agenda for the city. Although the mayor's leadership of the regime has been constantly under attack for concentrating its resources downtown at the expense of local neighborhoods, the mayor's popularity has kept his opponents from stopping his proposals and preventing his reelection.

As the regime prepares for the 1990s, it finds itself in a transition crisis. Young has won reelection for an unprecedented fifth term, primarily because he has maintained the confidence of his core constituency—black working-class voters—and the support of the business community. Enjoying the advantages of incumbency (project identification, campaign funds, and name recognition), Mayor Young has not been challenged seriously in any of his last four campaigns. Thomas Barrow, his opponent in the last two elections, was able to increase the percentage of anti-Young voters from 39% to 44%, but 51% is necessary to win in the city's at-large election system.

Nevertheless, the mayor continues to lose support among white voters. Some of the new anti-Young voters can be characterized as "he has been in there too long" types. Compared with their counterpart ABC voters (Anybody but Coleman), they were casting a vote *against* Young rather than *for* his opponent.

Although Young won the 1989 election by a comfortable margin, he felt the surge of the younger voters in polling booths. The 1990 Census will reveal a substantial buildup of 18- to 24-year-olds in the population, posing a serious problem for the governing regime. New 18-year-old voters were 3 years old when Young was first elected and so do not share the history of Young's veteran supporters. Having experienced no automobile boom or a major strike by workers, union leadership means little to them. The civil rights struggle also is history for them. They want a change in leadership, but so far they have not found a politician capable of articulating their frustrations and simultaneously appealing to their elders.

In preparation for the 1989 election, the mayor accumulated a multimillion dollar reelection campaign chest and secured most, if not all, endorsements. Even the Detroit Police Officers' Association, a longtime opponent of the mayor, endorsed Young. Young made a vigorous defense of his administration in the 1989 mayoral election. President of the City Council, Erma Henderson, the second most popular politician in the city, and Congressman John Conyers, a 25-year veteran of Congress, failed to survive the mayoral primary. In the runoff election, Thomas Barrow, the mayor's 1985 opponent, provided a replay of that election. Barrow did not create a stronger campaign organization or raise enough money and, as in 1985, Young was able to charge that Barrow was the suburbs' and the media's choice for mayor. Barrow was unable to convince the governing regime that he was a credible replacement for Young.

The reelection of Coleman Young does not mean that Detroit is set for the 1990s or that the mayor can revisit his failed initiatives of the 1980s. Many voters assume that this term could be his last, and the issue of his successor is unresolved. Some critics have raised the mayor's age of 72 as a factor, while others believe his staff is no longer up to the economic challenge of the city. Still others are dissatisfied with the pace of the mayor's attempts at replacing industry for the city. Yet there remains rather substantial support for the mayor, support that has become more emotional since the mayor has been under siege by local reporters.

The ongoing police scandal has created more anguish as more names and careers are entangled in the investigation. A recent newspaper poll revealed widespread uncertainty in the black community about the scandal. Former civilian deputy police chief Kenneth A. Weiner has been accused by the FBI of diverting drug funds from a secret police account to his dummy corporations located all over the nation. Weiner was linked to the mayor after he allegedly sold South African Krugerrands and conducted business with the mayor's private attorney. According to newspaper accounts, Weiner also paid the rent of the police chief's daughter's apartment in Los Angeles. The Weiner trials are still pending, but there has been no end to speculation. Whatever the outcome of this scandal, the reality is that the Detroit Police Department has suffered another blow to its reputation, diverting attention from the social and economic problems of the city.

CONCLUSION

The rise of black politics in Detroit has made members of the coalition more sensitive to race relations. In interviews with prominent members of this group, I found them concerned about the business and social image of Detroit (Rich, 1990). They support the work of New Detroit, Inc., and its advocacy of more services for the poor. They contribute to the upkeep of the Urban League and NAACP. Black politicians in the coalition have been tireless advocates of redistributive policies, but their efforts have been offset by repeated downturns of the city economy, giving Detroit less and less wealth to redistribute. Indeed, the economy has been so depressed that it has damaged the city's bargaining positions with its employee union and with developers.

Blacks' influence stops at the city limits. Detroit is not the only city in southeast Michigan chasing after the economic development buck, and no serious student of Detroit politics believes that black politicians are incorporated completely into the overall metro Detroit development strategy. The contiguous Oakland and McComb counties have promoted and nurtured the development of their urban areas at the expense of Detroit. In addition, Detroit still must compete with other Midwest cities for jobs, economic resources, and people. Detroit is very much a product of the marketplace, and lately it has not been able to compete effectively.

It is probably too early to predict that recent economic developments in downtown Detroit constitute a harbinger of recovery, but it is no exaggeration to say that the last decade has witnessed profound changes that seem to forecast more extensive developments in the near future. The 1980s were full of uneven development and uncoordinated construction. The process generated controversies, conflict, and compromises.

The preliminary figure for the 1990 Census, although challenged by the city, shows substantial outmigration of whites and blacks from the city of

Detroit. As a result, the city may lose a Congressperson or share one with the suburbs. The biggest loss will be in the state assembly, where Detroit could lose up to four seats. In the senate it could lose one seat. The loss of population will also affect the federal funds, now approximately $290 million, coming into the city. On a positive note, by maintaining a population of more than 1 million, Detroit continues to qualify for extensive state legislation created especially for the city. That legislation (33 separate laws) was limited to Detroit by specifying cities of more than 1 million, with Detroit always being the only such city in Michigan.

The greatest loss, however, will be of the black middle class, particularly the professional class. Aside from losing them as role models, Detroit will be losing them as the source for Detroit's civic leadership. They have been drawn to the suburbs by the fact that Detroit's property taxes, quality of schools, and safety conditions are not competitive with the suburbs.

The 1990s promise to continue the debate over the so-called downtown versus neighborhood rehabilitation strategies. Claims that the mayor has neglected the neighborhoods in his revitalization schemes will gather new support as President Bush has offered few resources for city rebuilding. Eventually, there simply will not be the class mix to make the city an interesting place to live.

Granted there is an element of rationality in Mayor Young's behavior, there is also a heavy element of romanticism in his projects and deals. He wants to be the Great Restorer of the city's glory days as one of the nation's most significant economic centers. Not only is he frustrated by inertia, but also by nay-sayers who believe it can't be done. The mayor quickly has embraced the new service economy based on the convention/tourism industry rather than romanticizing about the recovery of the automobile industry's share of the car market. Despite the political implications of gentrification, Young had supported the development of riverfront property. Accordingly, it is possible to argue that Detroit represents a combination of the market theory or city as maximizer suggested by Peterson (1981) and the regime theory of development or city development as social production espoused by Stone (1988). The governing regime of the city is not as diverse as Chicago's or as organized as Atlanta's. Nevertheless, Detroit is a laboratory of a regime in transition.

The restoration of the Detroit economy, a Herculean task because of the changing nature of the national economy, is made more difficult by the lack of cooperation with the suburbs. "Young bashing" has taken on a life of its own. Young has helped the situation as he often responded with ill-timed and intemperate comments. No politician or agency has attempted to mediate this conflict, making open warfare the norm. Detroit politics now is very different from the politics Banfield called a balancing act.

Although the mayor does not face an immediate election, he is not out of the woods on the crime issue. Crime remains Mayor Young's Achilles heel. Although Detroit has lost its title of "murder capital of America" to Washington, DC, it still has a serious crime control problem. Drug-related crime, particularly street crime, threatens to undermine the mayor's effort to sell the city as a safe place in which to invest. So far, the mayor has resisted more military-style enforcement policies and the inclination to violate civil liberties. President Bush has taken some initiative in the so-called drug war with a $9 billion budget, but the effort does not seem to be working; and the rhetoric of William Bennett, the former national drug czar, has not diminished the demand for drugs. If these drug enforcement initiatives fail, then cities like Detroit will be cities of war refugees.

In 1965 Banfield observed that, "Detroit mayors have so much power and receive so much attention that they tend, some say, to develop Napoleonic complexes" (pp. 53-54). His role models were Louis Miriani and Jerome Cavanagh, neither of whom was able to dominate the city politics in the manner of Coleman Young. The secret of Young's success as mayor may be his ability to take risks and to induce investors to take risks. He also is blessed with a less diverse population than was the case 25 years ago—few whites and fewer middle-class residents. In this political environment, political symbols and personality remain crucial to a politician's success. Young's personality is not so much Napoleonic as it is eudaemonistic. He enjoys taking risks. Operating best in the vast uncertainty that characterizes Detroit politics between projects, Young seems perfectly suited for this city's politics. Opponents to his developmental policy have failed, in part, because they simply were reacting to his proposals. None of the ad hoc opposition groups have offered any alternative plans of their own.

Ironically, the characteristics that made Young so popular with his electoral constituency also serve to alienate the support he badly needs from his policy constituency. In recent years scandal after scandal has drained energy from the mayor and his staff. When the administration spends time defending itself, policy stagnation takes over. The Young administration works best in an offensive rather than a defensive mode. As Young peers Robert Battle died and Congressman George Crockett retired, the mayor was left with fewer individuals willing to offer him a candid assessment of his administration's performance.

The preparation for a post-Young era is perhaps the tragedy of the 1990s. It is not clear whether the mayor will attempt a sixth term or who will succeed him. All the names mentioned share none of Young's history, intelligence, or special touch with his constituency. Without Young's dominant personality, black politicians may be inclined to make separate and self-serving deals with the business elite. For this reason some reformers are calling for a total restructuring of the city politics, ranging from metropolitanization to another charter revision.

Perhaps a return to the ward system is indicated. The subject of a petition drive for the coming 1990 election, a ward system would divide the city into nine districts. In 1941 the first attempt was made to change the system, and in 1954 it was proposed again by veteran State Senator Jackie Vaughn and rejected. A recent legislative proposal was defeated in the State House. Critics of the ward system have argued successfully that a ward system would generate corruption. Recently their argument has switched to the charge that a ward system would increase the mayor's power. The real reason the system has lasted so long is that the political brokers have wanted it that way, because it is easier for them to exert influence in this system. A balkanized system would be more difficult to manage and would result in the replacement of the most pliant white members. On the other hand, a ward system would be more representative and allow more accountability to individual constituencies.

Banfield concludes his 1965 review of Detroit politics with this observation: "Detroiters, voters as well as civic leaders, are aware that they live in a city packed with dynamite, and they do not play with matches" (Banfield, 1965, p. 63). In the summer of 1967 someone struck a match and the city blew up. The Cavanagh coalition collapsed and the city has never fully recovered. The Gribbs era was simply a transitional government.

It took most of Coleman Young's first term to create a new working coalition. This entrepreneurial-style, mayor-centered coalition is extremely effective at goal setting, but once a project is reified, the conflict begins. Success or failure of a project to generate jobs or revenue is ignored as a new set of goals is promulgated. What Stone and Sanders said about New Haven also holds for Detroit development politics: "The move from planning to execution was not a move characterized by a gathering momentum; rather, it was one of increasing conflict and disputed priorities" (Stone & Sanders, 1987, p. 179).

Stone offers a social-production model in which the business community works with the political community for the mutual benefit of the city. He concluded that:

> Regimes involve arrangements through which elements of the community are engaged in producing publicly significant results and providing a variety of small opportunities. The latter task often overshadows broader questions and makes it possible for governing coalitions to gain cooperation even though their larger goals enjoy only weak or uneven popular support (Stone, 1989, p. 235).

Although the small opportunities in Detroit are different from those in Atlanta, there is a tendency to go along and get along with the business community. Cooperation is achieved by inclusion of the black middle class in project development. The New Detroit, Inc., has been a vehicle for this cooperation. Businesspersons also contribute to black community organizations, churches, and campaigns. Everyone seems aware of the need for more white investments and involvement in the revitalization of the city, but no one has found a way to promote cooperation between the suburbs and the city.

In his 1989 victory speech Mayor Young held out the olive branch to suburbanites. Only a few mayors responded publicly. Most suburban mayors do not want a rapprochement with Detroit's officials because Coleman Young's problems serve as too useful a foil for their local politics. Nevertheless, it is clear that some type of metropolitan planning and cooperation is necessary if southeast Michigan hopes to avoid cannibalism by contiguous communities. Many suburban residents have been told that the faster the core city decays, the faster the suburban crescent will prosper. The truth is blight spreads; the faster it spreads, the wider it spreads.

REFERENCES

Agocs, C. (1974). Ethnicity in Detroit. In David W. Hartman (Ed.), *Immigrants and migrants: The Detroit ethnic experience* (pp. 390-408). Detroit: Wayne State University Press.
Banfield, E. C. (1965). *Big city politics*. New York: Random House.
Bledsoe, T. (1990). *From one world, three: Political changes in metropolitan Detroit*. Detroit: Center for Urban Studies, Wayne State University Press.
Conot, R. (1974). *American odyssey*. New York: William Morrow.
Deck, C. (1990, March 29). Detroit image hurts: Survey. *Detroit Free Press*, p. 1.
Eisinger, P. (1980). *The politics of displacement*. New York: Academic Press.
Jones, B., & Bachelor, L. (1986). *The sustaining hand*. Lawrence, KS: University of Kansas Press.
Laumann, E. O. (1969). The social structure of religious and ethnoreligious groups in a metropolitan community. *American Sociological Review, 34,* 182-197.
Peterson, P. (1981). *City limits*. Chicago: University of Chicago Press.
Rich, W. C. (1989). *Coleman Young and Detroit po ι.cs*. Detroit: Wayne State University Press.
Rich, W. C. (1990). The politics of casino gambling: Detroit style. *Urban Affairs Quarterly, 26,* 274-298.
Stone, C. (1988). Preemptive power: Floyd Hunter's community power structure reconsidered. *American Journal of Political Science, 32,* 82-104.
Stone, C. (1989). *Regime politics: Governing Atlanta 1946-1988*. Lawrence, KS: University of Kansas Press.
Stone, C., & Sanders, H. T. (Eds.). (1987). *The politics of urban development*. Lawrence, KS: University of Kansas Press.
Swanstrom, T. (1985). *The crisis of growth politics*. Philadelphia: Temple University Press.

6

St. Louis:
Racial Transition and Economic Development

ANDREW D. GLASSBERG

RACIAL CHANGE AND RACIAL CONFLICT

The politics of the city of St. Louis continue to revolve around the twin issues of population decline and race. In the 1970s, St. Louis suffered the largest percentage population decline of any large American city. That 27% population loss brought its population down to 453,000, just over half of the city's population at its high point, 856,000 in 1950, according to the Census. The 1980 population total was less than at any other time in the twentieth century.

The 1980s continued the trend of the previous decade, although not at so rapid a rate. A census conducted in St. Louis in 1988 as part of the Census Bureau's test of procedures for 1990 showed a total population of 405,000 (Todd & Schlinkman, 1989), and preliminary figures from the 1990 Census report a figure of 397,000 as the city's population total ("Preliminary census total," 1990). Although these figures suggest that the city's rate of population loss was cut in half in the 1980s, from 27% to 13%, this 13% rate was exceeded only by Detroit and Pittsburgh among the nation's 50 largest cities in the 1980s ("Preliminary census total," 1990).

As the city's total population changed, the composition of the population changed as well. The 1970 Census (Todd & Schlinkman, 1989) showed the city was 41% black; the 1980 Census reported 45% black; the 1988 Test Census reported 46% black; and the 1990 Census reported 48% black. As with the absolute numbers, these racial composition figures show a continuing trend, although a slowing one. Since the city's black population is considerably younger than the white population, there is no reason to anticipate the development of a black voting majority anytime soon, but expectations that the city eventually will have such a majority play a major part in its politics.

The political importance of race is increased by the extreme segregation of the city's population (one study reported it as among the most "hypersegregated" cities in the country) (Todd, 1989) and by the relative absence of significant populations of other minority groups. In the 1988 Test Census, blacks and whites together accounted for 98.6% of the city's population. This circumstance, now becoming rare among the nation's largest cities, makes the politics of race starker in St. Louis than elsewhere, and less amenable to coalitional politics.

This gradual change in the city's composition is not yet of sufficient magnitude to produce a black majority in the electorate. Because the underlying trends have been operating for a sufficiently long time and in a sufficiently polarized electoral climate, it is reasonable to expect that the dominant majority in city politics is approaching a major change. Even when relationships between the city government and external business groups appear to be stable (and they have not always been so), below the surface is the recognition that these patterns could change significantly as demographic changes inexorably play themselves out in the city's political arenas.

Existing city leadership responds to these trends with efforts at economic development and at reversing population trends. The city invests significant effort, therefore, in activities that it hopes will persuade existing middle-class city residents to remain within the city and will attract middle-class suburbanites to relocate within the city. While these activities are conducted without overt racial content, they have obvious racial implications.

Racial conflict in St. Louis is made more pointed by the generally low educational level, both black and white, of the populations in the metropolitan area. The 1980 Census showed that, of the 15 largest metropolitan areas, St. Louis ranked 13th in percentage of its adult work force with a college degree, and 14th in percentage of its adult work force with any graduate education (Glassberg, 1984). Because the city of St. Louis comprises a relatively small part of its metropolitan area, middle-class suburban neighborhoods that would be within city limits in "typical" older central cities are outside the city's boundaries in the St. Louis case.

Movement to the suburbs is not an exclusively white phenomenon in St. Louis. Both the 1970s and 1980s saw extensive black movements into St. Louis County, and the 1990 Census shows a substantial growth in the number of black residents in the county. Since this outmigration is more heavily middle class than the black population as a whole, one consequence is a severe shortage of well-educated leadership. This migration, coupled with generally low educational levels in the metro area, exacerbates the politics of race.

FORMAL GOVERNING STRUCTURE

The fundamental structure of the St. Louis city government remains as it has been for decades—a partisan mayor and Board of Aldermen. The mayor, together with the separately elected president of the Board of Aldermen and the city controller, form the Board of Estimate and Apportionment, which has independent powers over the city budget and the approval of city contracts and bill paying. Each of these citywide officials is elected for a four-year term. In an overwhelmingly Democratic city, victory in the Democratic primary is normally tantamount to election, although there has been at least one close challenge from a "write-in" campaign for president of the Board of Aldermen.

Within this setting, the mayor remains by far the most influential official, although city controllers periodically have used the power of their office over the payment of bills as a source of independent political leverage. The mayor exercises direct control over most city departments, but there are major exceptions.

The city's police department is an agency of the state government, not the city. The governor appoints four of the five members of the Board of Police Commissioners, with the mayor serving as an ex-officio fifth member. In recent years, interpretations of a "tax-revolt" amendment to the state Constitution have given the city government additional leverage over police affairs.

The mayor and the city government do not directly control such "county" offices as collector of revenue, license collector, and recorder of deeds; these are headed by citywide elected officials and staffed by those elected on a patronage basis outside the city's normal civil service system.

Despite the city's enormous population decline, membership of the Board of Aldermen remains at 28, with aldermen elected from single-member wards for four-year terms. Wards now have very small populations, less than 15,000 each. Although Republicans occasionally have been able to elect one or two aldermen, they have not been serious challengers in the vast majority of the city's wards, and as with citywide offices, the Democratic primary usually decides election outcomes.

A STRUCTURED ANARCHY OF POLITICAL INFLUENCE

Any discussion of political influentials in St. Louis must place the mayor at the center, but the mayor's hold on this status is really ex-officio and now does not last longer than the individual's service in office. This is a reflection of the relatively anarchic structure of the city's political life, the absence of effective political-machine organizations capable of mobilizing beyond narrow neighborhood or ward boundaries, and the growing constraints of race.

Mayoral candidates must build their own individual coalitions of support by responding to micro-issues at the neighborhood level. It has become successively harder for white candidates to build such coalitions across racial lines, and black candidates for citywide offices have not made significant inroads in white neighborhoods.

The role of the press is somewhat diminished in city politics (in contrast to its greater power in suburban St. Louis County) by the position of the *St. Louis Post-Dispatch,* the one daily paper, as a liberal voice in a generally conservative city.

Within the political leadership of the black community is intense controversy and positioning for leadership, driven in part by expectations of gaining control of city politics in the not-too-distant future. Representative William Clay (D-St. Louis) remains the dominant political force in the black community, but other actors now vie for attention as well. The current city controller, Virvus Jones, who acquired his position in a "job swap" engineered by the mayor and described below, has become a particularly visible articulator of black interests in the city's political controversies.

Citywide electoral contests have reflected both the process of racial transition and its incomplete nature. In the first mayoral campaign of the current incumbent, Vincent Schoemehl, in 1981, his major theme in successfully defeating the then-incumbent James Conway in the Democratic primary was his pledge to reopen a city hospital in predominantly black North St. Louis that had been closed under the Conway administration. Not only was this hospital never reopened, but the remaining city hospital, on the city's Near South Side, also was closed during Mayor Schoemehl's first term in office. Nevertheless, when Mayor Schoemehl first ran for reelection in 1985, he had the support of Representative Clay and triumphed over weak and divided opposition (Bargmann, 1990).

As the mayor's second term neared its end in 1989, it was assumed that he would have strong black opposition. A black former city alderman (and challenger of Representative Clay for leadership of the black community), Michael Roberts, had almost won the Democratic nomination for president of the Board of Aldermen in 1987. After barely losing the position in the 1983 Democratic primary and then losing the general election in a "write-in" campaign with the use of stickers permitted, Roberts lost the 1987 primary by only 171 votes (according to the Board of Election Commissioners' tally) and by about 60 votes (according to a recount panel appointed as part of a federal court suit challenging the outcome) (Order, Roberts v. Wamser et al., 1987).

Roberts was a candidate for the mayoral nomination in 1989 and might have mounted a strong challenge, although the continuation of white voting majorities would have made this a clearly uphill race. Mayor Schoemehl was able to undercut this possibility successfully by arranging a "job swap" in which the black city auditor (an appointive position) and the white city controller (an elected position) switched jobs. The mayor then ran for a third term with the black official, Virvus Jones, as his successful running mate for controller. Although the mayor had engineered the political maneuvers that made Jones controller, in the period since this last citywide election Jones has emerged as a powerful antagonist of the mayor and the most visible black political figure in the city.

Resistance to these political shifts toward greater black influence is severe and currently is focused most intensely on the St. Louis School Board. Unlike the city's Board of Aldermen, elected from small wards on a partisan basis, the School Board is, by state law, elected entirely at-large and without partisan endorsement. Candidates supported by local offshoots of white citizens' councils came close to a majority on the Board. A bi-racial slate, with heavy support from the city's business community, turned back this challenge in April 1991.

In recent years, controversies between the city controller and the mayor, long a feature of city political life, have taken on strong racial aspects as the black city controller seeks to use the office's power as a point of leverage in the city's ongoing conflicts over distribution of jobs and contracts. These conflicts have become especially acute regarding such large-scale city projects as the expansion of the Convention Center and the city's related effort to build a new sports stadium to replace the National Football League team that left the city for Phoenix at the beginning of 1988.

St. Louis's relationship with the state of Missouri continues to be one in which state decisions impinge on city autonomy. The city's police department remains under state, rather than city, control, a situation that has prevailed since shortly after the Civil War (Stein, 1991). This long-standing control has been weakened somewhat by indirect effects of the state's Hancock Amendment, a state Constitutional amendment limiting state and local taxing authority, passed by referendum in the early 1980s. While the most commented-on and controversial provisions of this amendment lie with its requirements for referenda approval of virtually all state and local tax or fee increases, the amendment also provides that the state must fund new state-imposed mandates on local governments. As interpreted by the courts, this provision could require the state to pay for additional police costs imposed by the state on the city government. Prior to the passage of the Hancock Amendment, the state legislature could approve increases in police salaries, for example, and mandate that the city government pay the costs.

While the Hancock Amendment does not give the city government direct control of the police department and the mayor still remains only one ex-officio member of the Police Board largely appointed by the governor, the provisions of the amendment do increase the city government's bargaining power and do serve to increase somewhat the mayor's control over the police department.

State influence over the structure of city government also remains strongly felt in the continuation of so-called "county" offices in the city government. These separately elected, patronage-staffed offices are part of the city government but not under mayoral control. Efforts by Schoemehl to increase municipal control of these offices have been thwarted largely by state legislative and gubernatorial action to maintain the independence of these positions and keep their staffs out of either mayoral control or the municipal civil service system.

Although the 1980s saw a limited shift of functions from municipal to state funding (such as the city teacher's college), St. Louis is positioned poorly to use state resources as any significant substitute for declines in federal assistance. Missouri remains one of the lowest tax jurisdictions in the country (Advisory Commission on Intergovernmental Relations, 1986; Phares, 1990). Although periodic proposals arise for changing the state tax system, these remain unlikely to occur in any form that might benefit a central city such as St. Louis as long as statewide electoral contests remain firmly in conservative Republican control.

The city has more success in developing relationships with its suburbs, which invariably means St. Louis County, its autonomous neighbor to the west. The county, which does not include the city, is now approximately two and one-half times as large as the city, and county executives have become increasingly assertive in taking the position that, because of this disparity in populations, the county should be the dominant player in regional matters. Relationships between the city's mayor and the county executive have fluctuated over time, depending upon whether the top issues on the agenda are ones of economic development, over which regional compromises usually emerge, or distributional policies within the metropolitan area, which are usually more conflictual.

Conflicts arise over functions that have moved to a regional basis, such as mass transit. The city and county frequently squabble over the expenditure of regional subsidies for the bus system, and over the airport, which although located in the county is owned by the city.

The city gains some influence over county affairs through a peculiar aspect of the state's structural control over city government. Since the placement of the city's current boundaries in 1876 and the city's concurrent withdrawal from St. Louis County, the state Constitution has provided a procedure for revamping the configuration of political power in the region. A Board of Freeholders may be created by citizen petition in the city and county and requires signatures from voters totaling at least 3% of the number of votes cast in the previous gubernatorial election in the city and county. The county executive and the mayor then each appoint members, with the governor appointing one additional member. This body is empowered to draft proposals for regional reorganization, but these proposals then go to referendum in the city and county and require concurrent approval in each area in order to go into effect (Schmandt, Steinbicker, & Wendel, 1961).

Changes in the Missouri Supreme Court's rulings concerning incorporation of new municipalities and annexation by existing municipalities in St. Louis county have produced an unstable political environment in the county. The county is at risk of having virtually all currently unincorporated areas either formed into new municipalities or annexed by existing ones. It could be left with only isolated poverty areas under direct county control, and this is perceived by county officials to be an impossible administrative and fiscal outcome. In addition, parts of St. Louis County are incorporated into a series of very tiny municipalities (the county now has nearly 100 separate municipal governments within its boundaries), and many of these are fiscally starved and barely able to deliver any municipal services.

In the late 1980s, in response to these conditions, county leaders sought to create a new Board of Freeholders to attempt a comprehensive restructuring of the county. Because of the state's Constitutional provisions, this effort could not be undertaken without the participation of the city as well. After the first efforts of the Board of Freeholders came to naught when the U.S. Supreme Court ruled the requirement that freeholders be property owners violated the U.S. Constitution, a revised Board of Electors was formed. Although this body is not expected to produce major structural changes, it is one vehicle for the city to exercise influence over the structure of the region beyond its own boundaries.

Cooperation between city and county has proven possible in areas of cultural facilities. The city and county share a Zoo-Museum District, and the past decade has seen the addition of a science center, botanical gardens, and history museum to this district. Efforts to add the symphony to this roster were defeated by the voters in 1989.

ANARCHIC POLITICS AND ORGANIZED BUSINESS

What the formal political structure separates, the business community unites. In contrast with the extreme fragmentation of political institutions and political power, the St. Louis business community is quite well organized. Two interlocking groups stand at the apex of the structure: the Regional Commerce and Growth association (RCGA) and Civic Progress, Inc.

Civic Progress, Inc., is a small organization, with membership limited to chief executives of corporations headquartered in the St. Louis area. A few powerful "branch managers," such as the head of the CBS-owned dominant radio station, are also members of the group. The heads of the three major universities in the area are invited to participate. Significant civic campaigns reach "critical mass" when they are endorsed by Civic Progress, Inc.

The Regional Commerce and Growth Association serves as a regional Chamber of Commerce, promoting the economic-development interests of the region. The activities of RCGA are internally, as well as externally, focused. Thus, the organization's "Sold on St. Louis" campaign directs as much energy at persuading area residents of the region's virtues as it does in merchandising the region elsewhere. Both Civic Progress, Inc., and RCGA clearly take a regional focus, covering both city and county at a minimum. Both groups will work against any efforts of city or county political figures perceived to be too parochial.

Missouri law continues to give local governments extensive development powers, including the use of eminent domain for private redevelopment projects and the abatement of most property taxes. The use of eminent domain is controversial and requires the passage of a specific ordinance by the Board of Aldermen declaring the proposed redevelopment site as "blighted." Sufficient examples of the successful use of this authority have shown property owners that they cannot assume they could successfully hold out against redevelopment schemes favored by the city government.

The impact of the city's broad use of property tax abatement is controversial. Proponents argue that it has led to extensive redevelopment activities, especially downtown, and has produced an attractive city core that is an inducement to both tourist and suburban resident utilization of the city's downtown. This, it is argued, produces jobs, earnings, and sales tax receipts that more than make up for the costs incurred (Mandelker, Feder, & Collins, 1980). Critics emphasize that, under Missouri law, decisions about abatement are at the sole discretion of the local general-purpose government (in this case the city of St. Louis) but are binding on other taxing authorities as well. The St. Louis School Board, therefore, must also abate its property taxes when the city government decides to take this step but does not share in any revenue increases the city might obtain from the sales or earnings tax.

The 1980s saw a continuation of the city's successful efforts at downtown redevelopment, with the completion of major shopping/entertainment complexes at the downtown St. Louis Centre and in the Rouse Corporation's renewal of the Union Station complex, just west of the downtown core. New hotels and large office buildings have added to this sense of downtown vitality. Surveys of city and suburban residents indicate that this development has produced positive and improved images of the city's downtown core among both central city and suburban residents of the metropolitan area. Incumbent city officials, especially the mayor, take credit for this trend, and the mayor is using it in his campaign for the Missouri governorship, an office . not held by a St. Louis-area resident for over 50 years (Bargmann, 1990).

For major construction projects, city-county cooperation has proved to be achievable. In each instance, the dominant business groupings, RCGA and Civic Progress, Inc., put their seals of approval on the projects. The city and county have worked together to secure federal funding for the beginnings of a regional light-rail system, whose construction has now begun. Similarly, the city and county are working cooperatively toward the construction of a new downtown stadium and convention center expansion, after a period of controversy in which the county unsuccessfully sought to have a new stadium built on the county's northwestern periphery, far removed from the city of St. Louis.

EDUCATION: THE CUTTING EDGE OF CONTROVERSY

Within the city and the larger metropolitan area, the politics of education continues to impinge sharply on the St. Louis city government even though public education is not one of its formal responsibilities. Education in the city is the responsibility of the separately elected St. Louis City School Board.

The School Board and education politics have been troublesome for the St. Louis city government. The School Board has been continually short of funds, but most efforts to provide additional resources, especially for capital improvements, have been stymied by state requirements for extraordinary majorities for their approval. (State Constitutional requirements were changed recently to reduce the needed majority from 2/3 to 4/7, and the School Board has now been able to secure approval under this eased requirement. School Board elections have become a focus for white resistance to city demographic change, and very poor citizens' opinions of the quality of city schools are seen by city officials as a major impediment to their efforts to attract more middle-class residents and to prevent existing middle-class residents from leaving.

Education politics is complicated greatly by federal court supervision, a result of court findings that the School Board never met Constitutional requirements for full desegregation of the schools. A series of court cases culminated in a consent decree that said the state was required to pay most of the costs of a regional desegregation order which, while de facto mandatory on the School Board and most of the school districts in St. Louis County, is voluntary for individual pupils. As a result of this order, 12,000 city students attend public schools in the county, while only about 700 county students attend city public magnet schools (Confluence St. Louis Task Force on Racial Polarization, 1989).

ST. LOUIS IN THE 1990s

The economics of the St. Louis region have continued to be sluggish, with the twin bases of manufacturing economy and automobiles and defense production both operating in highly uncertain environments at the opening of the 1990s. The St. Louis region continues to do well as a headquarters location, however, as the area benefits from the combination of its central location and low cost of living.

These trends, combined with continued city use of its abatement powers, have produced extensive downtown redevelopment, as well as some redevelopment of the city's extensive historic housing stock, but changes in tax incentives for historic redevelopment resulting from the federal Tax Reform Act of 1986 have considerably reduced this type of development in recent years.

St. Louis has been a city highly dependent upon federal programs (Schmandt, Wendel, & Tomey, 1983), in part because of the relatively small level of state involvement. As some of these federal initiatives decline, and as defense production becomes more problematic, the economic future of the region looks cloudy to its leaders. It remains to be seen what impact this situation will have on the political structure of the city. While the region's economic leaders continue to advocate regional restructuring in response to changing conditions, it can be expected that restructuring will meet substantial resistance within the city, especially as the city moves ever nearer toward a change in the racial composition of the city's voting majority.

Even though the past several decades have been characterized by city governments responsive to business development and willing to promote additional economic development, it is questionable whether this stable structure will survive racial transition in the city's elected leadership. A major political challenge for the city through the 1990s will be determining a pattern of leadership responsive to the dual facts of black voting majorities and a cohesive business community. Whether this leadership will be cooperative or adversarial remains to be seen.

Whether a regime (Stone, 1989) based on a strong and organized business community and a highly fragmented political community would survive a change in electoral democracy in the city toward a black voting majority remains unknown. The current governing regime of St. Louis, therefore, might best be characterized as stable but brittle and susceptible to change as the city itself changes.

REFERENCES

Advisory Commission on Intergovernmental Relations. (1986). *State fiscal capacity and effort.* Washington, DC: Government Printing Office.

Bargmann, J. (1990, September). What makes Vince run? *St. Louis Magazine,* pp. 50-57, 95-96.

Confluence St. Louis Task Force on Racial Polarization in the St. Louis Metropolitan Area. 1989. *A new spirit for St. Louis: Valuing diversity.* St. Louis: Author.

Glassberg, A. (1984, September). *Graduate education rates by metropolitan area.* Report to Dean of the Graduate School, University of Missouri-St. Louis, St. Louis, MO.

Mandelker, D. R., Feder, G., & Collins, M. P. (1980). *Reviving cities with tax abatement.* New Brunswick, NJ: Rutgers University Center for Urban Policy Research.

Order, Roberts v. Wamser et al. (E.D. Missouri, 1987).

Phares, D. (1990). Missouri well off but stingy. *St. Louis Journalism Review, 20*(129), 1, 6-7.

Preliminary census total: 245,837,683. (1990, August 30). *USA Today,* p.8A.

Schmandt, H. J., Steinbicker, P. G., & Wendel, G. D. (1961). *Metropolitan reform in St. Louis.* New York: Holt, Rinehart & Winston.

Schmandt, H. J., Wendel, G. D., & Tomey, E. A. (1983). *Federal aid to St. Louis.* Washington: Brookings Institution.

Stein, L. (1991). *Holding bureaucrats responsible.* Tuscaloosa: University of Alabama Press.

Stone, C. (1989). *Regime politics.* Lawrence, KS: University Press of Kansas.

Todd, C. (1989, August 4). St. Louis called "hyper-segregated." *St. Louis Post Dispatch,* p. 1.

Todd, C., & Schlinkman, M. (1989, August 4). Test census results show racial changes. *St. Louis Post Dispatch,* p. 23A.

7

Atlanta: Urban Coalitions in a Suburban Sea

ARNOLD FLEISCHMANN

THE DEMOGRAPHIC AND ECONOMIC CONTEXT OF ATLANTA POLITICS

POPULATION CHANGE

Several demographic changes characterize Atlanta since 1960: an overall loss of residents, a dwindling share of the metropolitan area's population, the increasing proportion of black residents, and the lingering presence of poverty and other social problems. Like other large cities, especially those in the Northeast and Midwest, Atlanta has not experienced any appreciable population growth for several decades. An increase of less than 10,000 residents occurred between 1960 and 1970, when the city's population reached its peak of 495,039. Since then the total has fallen to roughly 394,000 in 1990, a drop of 20%.

While Atlanta has lost population, its suburbs have gained substantially. From 1980 to 1988 the metropolitan area registered a gain of 600,000 residents, including 50,000 in Fulton County and 60,000 in DeKalb County outside the Atlanta city limits.[1] Growth was even more dramatic in two affluent suburban counties: Cobb grew by 127,000 residents and Gwinnett by 156,000 (Bachtel, 1990, pp. 123-140). One consequence of these changes is Atlanta's diminishing significance within the metropolitan area.[2] In 1960 the city accounted for 41.7% of the metropolitan population. By 1970 the ratio had dropped to 31%. In 1988 it represented only 15% of the area's 2.7 million residents (U.S. Bureau of the Census, 1990c, 1990d).

Perhaps of greater significance than Atlanta's loss of residents were changes in the composition of its population. Atlanta went from 38% black in 1960 to virtual racial parity in 1970. By 1985 the city was more than two

---

AUTHOR'S NOTE: My thanks to Chuck Bullock, Carol Pierannunzi, Hank Savitch, and Clarence Stone for their suggestions regarding an earlier draft of this chapter.

thirds black and included 55.8% of the black population in the seven core counties. Sizable numbers of blacks also have settled in Fulton County just south of Atlanta's city limits and in suburban DeKalb County, although these suburban areas have become overwhelmingly black, as have adjacent areas in the city (see Clark, 1988).

Atlanta shoulders a substantial part of the metropolitan area's social problems. In 1980, 48.5% of the metropolitan area households receiving public assistance were in Atlanta. Similarly, 27.5% of Atlanta's residents lived below the poverty level in 1979, compared with 12.2% for the metropolitan area as a whole. By 1987 the city's per capita income had risen to $11,689, which was $2,300-$4,000 below the surrounding metropolitan area (U.S. Bureau of the Census, 1983, 1990c). In addition, there is the danger of Atlanta becoming a bifurcated city of the college-educated and an "urban underclass," with few residents in between. Atlanta also has competed in recent years for the ignominious title of the nation's "murder capital" and had a serious crime rate in 1983 that was almost double the rate for the metropolitan area as a whole. The city has 50,000 public-housing tenants and has acquired a homeless population that may number 20,000 (U.S. Bureau of the Census, 1986; Feuerstein, 1990, [Metro] p. 6).

As one might expect, these population patterns have affected the metropolitan area's local tax bases. In 1966 Atlanta accounted for 89% of the real and personal property subject to taxation in Fulton County and 45% of the total for the metro area. Twenty years later, taxable property in Atlanta was assessed at over $6 billion, which was only 54.7% of the Fulton County tax base and 17.5% of the total for the 18 counties in the MSA (U.S. Bureau of the Census, 1968, 1989a). This suburban growth has been associated with major expansion of retail and employment centers and growing suburban clout in statewide elections.

THE LOCAL ECONOMY

Atlanta's economy has displayed both continuities and changes during the past 30 years. Two constants have characterized the local economy: its economic base and growth. Unlike the South and the rest of Georgia in 1960, the Atlanta metropolitan area was linked already to the service economy: 24.1% of the workers identified by industry in the Census were employed in services, as were 28.9% in the city of Atlanta. Another 22.4% of metropolitan workers were in wholesale and retail, as were 23.5% of Atlantans. In both the city and the metropolitan area, about 7% of workers were in finance, insurance, and real estate; slightly less than 10% were employed in transportation, utilities, and communications. The figures for manufacturing were 23.2% in the metropolitan area and 19.2% in the city, but 27.1% in the state (U.S. Bureau of the Census, 1963, 1964).

Atlanta's reliance upon commercial and service employment has accelerated since 1960. By 1980, manufacturing accounted for only 17% of total employment in the metropolitan area and 13.2% in Atlanta. This compares with 24.1% for Georgia. Service employment was 28.3% in Atlanta and 37.8% in the MSA, compared with 26.6% in the state. The city's relative share of Georgia's work force has diminished, however, from 14.2% in 1960 to 7.4% in 1980. In contrast, the metro area rose from 28.5% of Georgia's workers in 1960 to 41.4% 20 years later (U.S. Bureau of the Census, 1983).

The other mainstay of the Atlanta economy has been its growth, which generally has exceeded national and state levels during the past 30 years. This was especially true during the economic upheaval of the early 1980s. From 1977 to 1982, manufacturing employment nationally dropped 2.5%, but it rose 14% in metropolitan Atlanta. During the same five years, national retail sales expanded 47.4%, compared with 65% in the Atlanta MSA.

A major change in Atlanta's economy has been its transformation as a corporate center. The city was virtually invisible as a headquarters town in 1965. None of *Fortune*'s 500 largest industrial firms were headquartered in the Atlanta area. Even Coca-Cola had its corporate offices in New York City, although Atlanta did have one each of the largest banks, merchandising firms, transportation companies, and utilities listed by *Fortune* ("*Fortune* Directory," 1965; "*Fortune* Directory II," 1965). By 1990, relocation and the growth of existing firms left Atlanta with 21 locally headquartered companies on *Fortune*'s list: 9 of the 500 largest industrials (including Georgia-Pacific and Coca-Cola), 4 utilities, 3 banks, 2 diversified-service companies, 2 transportation companies, and 1 retailer ("*Fortune* 500," 1990; "*Fortune* 500 Service Corporations," 1990).

Changes in Atlanta's role as a business center are evident in the city's growing international ties. By 1990, consulates for 39 countries, offices of 2 Canadian provinces, and such groups as the Australian, German American, and Japanese Chambers of Commerce had been established. Major British, Canadian, Dutch, German, and Japanese banks maintain agencies or representatives in Atlanta, primarily for investment banking (Cline, 1987; Bueno, 1990). It has been estimated that since the early 1970s, foreign investment in Georgia has totalled $1.8 billion from Canada, $1.75 billion from Japan, and roughly $1 billion each from the United Kingdom and the Netherlands. Japanese investors are noted especially for buying or building highly visible Atlanta office towers. Another indicator of Atlanta's rising international prominence is Japanese Prime Minister Kaifu's visit to the city in July 1990—his first stop following the summit in Houston for the leaders of the world's major democracies (Saporta, 1990b). Based on all of these trends, *Fortune* (Labich, 1989) designated Atlanta one of the nation's two "best cities for business."

POLITICAL CHANGE, 1960-1990

GOVERNMENT STRUCTURE

The most apparent changes in Atlanta's governmental structure occurred with the 1973 charter, which manifests the recognition by Atlanta's leaders that the city would have a permanent black majority. A Charter Commission appointed by the Georgia General Assembly drafted a document replacing the at-large Board of Aldermen with a mixed system of 12 district and 6 at-large members. Although black legislators influenced the composition of the Charter Commission and the final document, district seats guaranteed whites more representation than they would have secured under an at-large system. Neighborhood activism during the period also influenced adoption of district elections and mandatory citizen participation in city planning.

Like its predecessor, the new charter provided for a nonpartisan ballot and four-year terms beginning with the 1973 election. It granted the mayor veto and broad organizational powers, although it did mandate the existence of the law and finance departments. The document established a Council president elected citywide who would preside over Council meetings and have broad legislative powers, especially in terms of committees. The charter created a school board with six district and three at-large seats (Jones, 1978; Stone, 1989).

The grant of mayoral power in the new charter led to major changes that were not necessarily permanent. Both mayors Maynard Jackson and Andrew Young made extensive use of the ability to reorganize agencies. Jackson created a Public Safety Commission to increase his control over the police department (Stone, 1989). Interestingly, he also abolished the Commission after returning to the mayor's office in 1990 as a way of removing the public-safety director appointed by Andrew Young. Young, in turn, had reorganized positions into a three-person management team that removed him from day-to-day management but increased his staff's control over departments (Englade, 1985).

A less visible structural change is the pattern common in many cities of assigning functions to independent authorities. Several such bodies have been created by the state. The Metropolitan Atlanta Rapid Transit Authority (MARTA), which experienced some early political setbacks, was adopted in a 1971 referendum in Atlanta and Fulton and DeKalb counties but rejected by voters in Cobb and Gwinnett counties. With its own board and guaranteed sales tax, MARTA operates at arm's length from the local governments in its service area. The board is not isolated from political pressure, however, especially regarding fares and subway routes (Stone, 1989). The state has also created the Georgia World Congress Center Authority, which owns and operates the major convention center and new domed stadium in downtown Atlanta; the Downtown Development Authority, which issues revenue bonds for public and private development projects; and the Metropolitan Atlanta Olympic Games Authority, which has been granted broad fiscal and eminent-domain powers as part of its mission to stage the 1996 Summer Olympics.

THE DEVELOPMENT OF BLACK ELECTORAL POWER

Blacks have been key players in Atlanta's politics since the 1940s. Georgia's white primary was eliminated in 1946, and blacks constituted 40% of the city's registered voters as early as 1948. For most of the 1950s and 1960s, this bloc was highly cohesive and readily mobilized by the Atlanta Negro Voters League, which was controlled by black businessmen, ministers, lawyers, and college professors (Hornsby, 1977).

The 1949 reelection of Mayor William B. Hartsfield cemented a coalition between blacks and middle-class whites that excluded poor whites, the staunchest supporters of segregation. This was in marked contrast to other large cities in the South, where white politicians seeking municipal offices tended either to unite white voters in opposition to black-supported candidates or to build coalitions along class rather than racial lines (Murray & Vedlitz, 1978). White business leaders recognized, as Banfield noted, that "Atlanta must come to terms with the Negro if it is to continue to grow and prosper" (Banfield, 1965, p. 30; also see Allen, 1971, pp. 87, 107).

Hartsfield was succeeded by Ivan Allen, Jr., Chamber of Commerce president and candidate of the local "power structure." Allen outmaneuvered the other moderate mayoral contender in 1961 by attacking segregationist Lester Maddox, whom he beat in a runoff thanks to the coalition of blacks and middle-class whites. Considered progressive on racial matters, Allen was especially vilified for his 1963 congressional testimony in favor of a federal civil rights bill. Allen also served as a link between older, established black leaders and downtown business interests. Along with Allen's landslide reelection in 1965, blacks gained a seat on the Board of Aldermen for the first time since the nineteenth century (Hornsby, 1977; Allen, 1971).

Black representation increased during the 1960s although whites still dominated municipal government. The 1969 election witnessed the first major splits in the old electoral coalition. Generational differences had already emerged among blacks, and when Allen decided not to seek reelection, divisions also occurred within the white business community. The field had four major candidates, including one black, Dr. Horace Tate. Because of disagreements over possible scenarios and Tate's viability, black leaders divided their support. Sam Massell was elected mayor with the overwhelming backing of the black electorate in a runoff against the business elite's candidate. Maynard Jackson was the first black to win the citywide post of vice-mayor, but blacks won only 4 of 18 seats on the Board of Aldermen (Hornsby, 1977; Jones, 1978; Allen, 1971).

Even though Atlanta achieved a black majority by 1970, the Board of Aldermen still was made up overwhelmingly of whites. Black dissatisfaction with Massell grew, and Maynard Jackson challenged him for mayor in the 1973 election, the first under the new charter. Voting in the general election was generally along racial lines although black precincts were more likely to support white candidates than the other way around. Higher turnout among blacks than whites helped Jackson, but blacks' greater likelihood of skipping lower-level contests on the ballot hurt other black candidates. Massell and Jackson were forced into a runoff in which race became a key factor although Jackson benefited more than Massell from racial crossover voting. In the runoff, Jackson was elected mayor and a white, Wyche Fowler, was elected City Council president. The result was a Council split evenly between blacks and whites, a white as presiding officer, and the city's first black mayor (Hornsby, 1977; Jones, 1978).

Blacks have held the Council presidency since Wyche Fowler resigned in 1977. They also have occupied a majority of both the district and at-large Council seats since 1978. Jackson was reelected in 1977, following stormy relations with the business community that were repaired somewhat during his second term. He was not allowed to serve a third term, however, and 1981 began to shape up as another pitched battle (Stone, 1989). The preeminent black candidate was Andrew Young, a former member of Congress and ambassador to the United Nations. The leading white contender was state representative Sidney Marcus, considered progressive but the choice of the city's business elite. Young and Marcus appealed to white and black voters. Although he led Marcus by only 2% in the first election, Young took the runoff with 55% of the vote. Race and issues figured in the outcome, but only a limited percentage voted for a candidate of their own race who held different positions on issues (Bullock & Campbell, 1984).

Young was reelected handily in 1985 despite concerns about his detached management style, complaints over frequent traveling, and opposition from neighborhood activists. Young's second term cemented the strong ties he had built with downtown business. He continued his support of the Presidential Parkway and tried to recruit international firms. Especially visible were his efforts to promote downtown office construction and tourism, which included a partnership with the Rouse Company to redevelop Underground Atlanta as an entertainment complex, support for a new domed stadium, and successful bids to host the 1988 Democratic National Convention and 1996 Olympics. These efforts, along with such previous actions as support for a

piggyback rail yard in a poor white area and reduced support for neighborhood planning, further alienated Young from the neighborhood movement,
which had limited political clout by the end of the 1980s (Stone, 1989;
Englade, 1985).

Andrew Young was prohibited from seeking reelection in 1989. A number
of politicians jockeyed to succeed him, although there were few expectations
that a white could be elected. The contest shaped up as one between two
high-profile blacks, former mayor Maynard Jackson and Fulton County
Commission chairman Michael Lomax. Although Lomax benefited from
early financial support by business, especially development interests, he was
unable to dent Jackson's firm lead in the polls and abandoned his campaign.
Although City Council member Hosea Williams, regarded as an advocate for
Atlanta's poor, entered the mayoral race, Jackson won with nearly 80% of
the vote and a campaign that addressed the concerns of middle-income blacks
(Pierannunzi & Hutcheson, 1990; Sack & Copeland, 1989).

Jackson resumed the mayor's job in January 1990. Unlike his earlier
tenure, blacks controlled two thirds of the City Council seats. Almost all of
the blacks and whites on the Council are professionals such as lawyers,
business owners, and real estate agents. Most have strong ties, at least with
campaign financing, to the business community (Sherman, 1990; Thomas,
1990). Under such conditions and the continuous pressure to promote
growth, black control of political institutions does not translate automatically
into the politics of redistribution (Stone, 1989).

THE WIDER CONTEXT OF ATLANTA'S POLITICS

While black electoral power transformed Atlanta politics in many ways,
the city was becoming only one player in a larger governmental system. A
generation ago, Banfield concluded that "Fulton County government clearly
is overshadowed by the Atlanta city government" (Banfield, 1965, pp. 21-
22). In one sense, he was quite correct—the county spent $37.2 million, while
the city expended over $67.6 million in 1966—because other governments
had few residents and limited services. Conditions have changed, however.
By fiscal 1986-1987, local governments in the MSA spent more than $5.9
billion. While Atlanta spent $586.7 million (9.9% of the total), most notice-
able was spending by the area's counties: $409.5 million by Fulton, $331
million by DeKalb, $225.6 million by Cobb, and nearly $173 million by
Gwinnett (U.S. Bureau of the Census, 1970, 1971, 1989b, 1990a, 1990b).

Atlanta retains the largest budget in the metropolitan area, but its declining
share of public spending reflects major changes in postwar government.
Growth in population and service demands have transformed government in
suburban counties. Their work forces have expanded and become more

professional. They furnish police and fire protection; maintain water, sewer, and sanitation systems; and regulate land use in unincorporated areas. One has started its own transit system. They also have public schools with much more positive images than Atlanta's. For most metropolitan area citizens, then, what goes on in Atlanta is irrelevant to the quality of public services they receive.

Of equal or greater importance to changes in local government has been burgeoning state spending, which affects Atlanta in two ways. One is the substantial infrastructure investment that aids growth in outlying areas. For instance, the seven-county transportation plan prepared by the Atlanta Regional Commission (ARC) projected highway costs in excess of $1.8 billion during 1987 to 1994 that included over $600 million in federal money and $458 million in state expenditures. Roughly half of each total went to so-called regional projects, such as widening or creating new interchanges on the interstate system. These plans included the area's second freeway bypass through the affluent northern suburbs. Of the nonregional federal project costs, $34.6 million were planned for Atlanta and $301.9 million for the remainder of the seven-county area. Georgia's $252.4 million in costs projected for 1987 to 1994 included only $21.5 million in Atlanta (Atlanta Regional Commission, 1987). Similar patterns are expected for highway spending early into the next century. Moreover, over 60% of the nearly $2 billion proposed for subway expansion covers parts of Fulton and DeKalb counties outside Atlanta and a $361 million extension to Gwinnett County, which is currently not part of MARTA (ARC, 1987).

The second way the state has affected Atlanta is through its control of facilities crucial to the city's future as a site for major meetings and events. The time is past when the city's "power structure" could build a stadium and attract a major league baseball team (see Allen, 1971). What might be considered a breakthrough in state government hostility to Atlanta occurred in 1972 when the legislature created the independent authority that runs the Georgia World Congress Center. The GWCC competes with New York, Chicago, and a handful of other cities for the nation's major trade shows and conventions. Attendance at GWCC events has exceeded 1 million since the mid-1980s, and the Authority claims an impact on the state economy of nearly $1 billion during fiscal 1988-1989. The GWCC, which normally earns a small profit, relies on revenue bonds to fund expansion and is guaranteed part of Atlanta's hotel/motel tax.

Along with the nearby Omni Arena, the GWCC played an integral part in hosting the 1988 Democratic National Convention. State officials also took the lead in having the GWCC build and manage a 70,000-seat domed stadium that also could be used for conventions and other events besides football. The

$210-million Georgia Dome is scheduled for a 1992 completion and has been selected as the site for the 1994 Super Bowl. It, along with an expanded GWCC convention facility, is integral to Atlanta's plans for the 1996 Olympic Games (Georgia World Congress Center, 1989).

In addition to facing the growing influence of other governments, Atlanta increasingly is adrift in a sea of Republican suburbs. Although most municipal elections in Georgia are nonpartisan, county offices are not, and Republicans held 32 of the 87 county commission seats (36.8%) in the metropolitan area by 1989. This success includes control of Cobb, Gwinnett, and three other commissions. The GOP had at least 1 seat in 10 other metropolitan counties (Georgia Secretary of State, 1989).

At the state level, legislative redistricting in the 1960s increased urban representation in the Georgia General Assembly, but rural interests kept substantial influence over the agenda by controlling key leadership and committee positions (Hawkins & Whelchel, 1968). By 1989, metropolitan Atlanta's 18 counties were represented by 71 members in the House (39.4% of the total) and 22 in the Senate (40%). Redistricting following the 1990 Census will increase suburban representation, which could approach half the seats in each chamber. Thus, as Atlanta's suburbs continue to add new residents while the city and traditionally Democratic parts of rural Georgia continue to lose population, Republican influence will increase in the Georgia General Assembly. This may pose an additional barrier to addressing Atlanta's social problems as partisanship accentuates long-standing rural and growing suburban hostility to Atlanta, which has been tied to racial politics (Key, 1949; Schwartzkopff & Yancey, 1990).

POLITICAL POWER AND REGIME CHANGE IN ATLANTA

When Banfield (1965) published *Big City Politics,* Atlanta's future could be shaped by a small group of men (Stone, 1989; Allen, 1971). Black electoral clout, however, has allowed control of City Hall to pass from white businessmen to a generation of black politicians who matured during the civil rights movement. White business leaders did not respond by abandoning Atlanta. Rather, they forged a new governing coalition with black mayors and City Council members. This transition was far from smooth, and while it did leave some lasting changes in the local policy agenda, the new regime also maintained the commitment of economic and political elites to promoting growth.

BLACK ELECTORAL POWER AND REGIME CHANGE

The regime that governed Atlanta immediately after World War II was like that in many other cities, with little difference between City Hall and the Chamber of Commerce. The governing coalition under Mayor Hartsfield was dominated by white business leaders and promoted reform-style municipal government. The power of this group was tempered by the size of the black electorate, whose cohesive and moderate leaders became "junior partners" in the coalition (Stone, 1989). Elite bargaining between white and black leaders involved little public participation or scrutiny, but it produced consensus on such issues as urban renewal and housing. The marginal changes in segregation during the Hartsfield years, which diverged from official actions in other Southern cities, earned Atlanta a progressive reputation and bolstered business leaders' efforts to promote growth (Stone, 1989).

During the 1960s the city promoted the same agenda Mayor Allen had advocated during his time as Chamber of Commerce president: school desegregation, a freeway system, downtown development, mass transit, and other physical improvements. The governing coalition also promoted reform-style political institutions and governmental efficiency. The regime was threatened, though, as a generation of less moderate and more impatient black leaders came on the political scene. Urban renewal and other development initiatives also engendered citizen resistance (Stone, 1976, 1989; Allen, 1971).

Sam Massell alienated blacks and left business leaders split as five candidates vied for the mayor's job in the 1973 election (Stone, 1989). Maynard Jackson's runoff win over Massell seemed to pose an immediate and direct threat to the old regime. Jackson supported increased minority representation in the municipal work force and among contractors on city projects, reform of the police department, and neighborhood involvement in planning. An early conflict was the mayor's insistence that 20% of contracts on the new airport go to minority firms. He also threatened to withdraw city funds from banks that would not open board and executive positions to women and minorities.

Subsequent actions, however, helped forge a governing coalition that extended into the 1990s. Jackson was rebuked early in a letter from the chairman of Central Atlanta Progress, the well-organized voice of downtown business interests. The widely publicized letter was especially critical of the mayor's inaccessibility to business. Jackson reached a compromise over airport contracting and increased his contacts with business leaders, in part because of his efforts to promote economic development. For their part, business leaders promoted public-private partnership and increased their efforts to maintain close contacts with City Council members. As a result of mutual interests, a governing coalition between black politicians and white business leaders was established by the time Jackson left office in 1982 (Stone, 1989; Henson & King, 1982).

Andrew Young was not the business community's candidate in the 1981 election. Nonetheless, he and several other black politicians solidified the regime that jelled under Jackson. The governing coalition was noted especially for its efforts to promote development, which often brought it into conflict with poor neighborhoods (Stone, 1989). The coalition and bargaining between white business and black politicians were sustained when Maynard Jackson was elected mayor again in November 1989 (see Banks, 1987).

THE RISE OF LOCAL INTEREST GROUPS

The 1970s and 1980s saw the proliferation of new political participants activated by territorial and sectoral interests (see Walker, 1988). The involvement of neighborhood and preservation groups in land-use planning, especially in low-income areas, increased after adoption of the new city charter. Moreover, community participation had a greater impact on zoning under the neighborhood planning system implemented in the mid-1970s (Hutcheson & Prather, 1988). These groups proved no match for the governing coalition over the long haul, however, given the shared resources and agendas of business groups and elected officials (Stone, 1989).

Business interests no longer speak with one voice in metropolitan Atlanta; they have become more active along geographical and sectoral lines. This does not necessarily pose a threat to such traditional groups as the daily newspapers, Chamber of Commerce, and Central Atlanta Progress, which have remained active in promoting the Georgia Dome, the Underground Atlanta entertainment complex, crime control, downtown housing, and similar programs (Saporta, 1990c, 1990a). Business groups have developed in certain sections of the city, as well as in the suburbs. Midtown and Buckhead have organizations with several hundred member firms, a full-time staff, and modest budgets to influence policies affecting their sections of the city. Their resources pale, however, compared with those of the Atlanta Chamber of Commerce and the Atlanta Convention and Visitors Bureau (Saporta, 1989).

Businesses generally advance their sectoral interests with campaign donations. Developers are active givers in City Council campaigns. The same is true of firms dealing with the city, including major minority contractors (Thomas, 1989; Sherman, 1990). Unlike the singular importance of the mayor's office in the 1960s, a wide range of business organizations now maintains ongoing contact with the legislative branch.

A PERMANENT COUNCIL?

The mayor and City Council always have been key players in Atlanta's politics, although the mayor has been lead actor. At least since the new city charter, however, Council members have become individual political entrepreneurs easily reelected every four years. For the seven quadrennial Council elections beginning in 1965, an average of 32.8% of the winners had not been on the Council four years earlier. The only majority turnover occurred in 1973 under the new city charter, when 11 of the 18 members were elected to their first term. Three other councils had 7 new members. These figures somewhat understate the power of incumbents, however. In 1969, for instance, 2 of the 7 new members were elected because the size of the Council was increased from 16 to 18 seats. One new face was seated on the 1986 Council only when John Lewis resigned after winning a seat in the U.S. House of Representatives.

The Council that took office in 1990 seems well insulated politically. Of the 13 members and Council president reelected, 1 each has served since 1962, 1970, 1974, 1980, and 1986; 3 joined the Council in 1982; and 6 began their service in 1978. Nine were elected without opposition. Although 5 members were not on the Council in 1986, 1 took the seat of an incumbent who ran for mayor in 1989, and another had already replaced a three-term incumbent convicted for his dealings with a developer.

The secret to this success seems to be three C's—cash, casework, and contacts. Council President Marvin Arrington, who served 10 years before winning his current post in 1980, raised $122,500 in campaign funds during 1989 even though he had no opponent. Seven Council members garnered over $50,000 in donations. Lists of contributors to Council campaigns resemble a "who's who" of Atlanta business. Especially prominent are firms linked to the city. Three developers, for instance, distributed a combined $66,700 to candidates in 1989, while three vendors at city-owned Hartsfield International Airport contributed more than $50,000. Although many donors spread their money among Council members, others target donations based on Council committee assignments (Sherman, 1990).

Council members have also developed an extensive system of constituent service. In addition to an annual salary of $22,500, each member is allowed $50,000 for individual staff aides, $5,000 for annual travel, and a mobile telephone. Members also have used their expertise in law, real estate, and various types of consulting to develop links with some firms that do business with the city. Thus, as Atlanta enters the 1990s, middle-class professionals on the City Council have nurtured a system that makes them electorally secure. Some critics argue that it enriches Council members and their friends (Thomas, 1989, 1990a; Hayes, 1983; Stone, 1989).

SOCIAL CHANGE, ECONOMIC DEVELOPMENT, AND PUBLIC POLICY

Atlanta's development has been linked to public policies in several ways. First, members of the governing coalition have not been passive observers of demographic trends. For instance, concentrating the black population south of downtown was not accidental or "natural." Even after racial zoning was ruled unconstitutional in the 1920s, highways, public housing, and urban renewal placed barriers between Atlanta's black South Side and the white North Side (Bayor, 1989). Government policy has been reinforced by private action, even after enforcement of racially restrictive covenants was outlawed in 1948. During the late 1980s, the location of financial institution branches, lenders' relationships with real estate agents, high rejection rates for black mortgage applicants, and restrictive lending to small businesses have been linked to the problems of neighborhood viability on Atlanta's South Side (Dedman, 1988a, 1988b, 1988c, 1988d).

Political and business leaders also responded to population changes by trying to expand the city's boundaries. In the late 1940s, boosters hoped the addition of new residents would allow them to brag about Atlanta's size compared with that of other cities. More territory also meant augmenting the tax base and politically leveraging the expanding black population with the addition of white middle-class voters. These efforts eventually led to the 1951 Plan of Improvement adopted with the support of city officials, the Chamber of Commerce, the Atlanta Negro Voters League, the local delegations in the Georgia General Assembly, and other groups. The Plan added 82 square miles and 100,000 people to the city in 1952. It also temporarily limited Fulton County's role as a service provider (Rice, 1981).

The territorial expansion of the 1950s had significant consequences for the makeup of Atlanta's population. The 33 census tracts in 1960 that most closely approximated the annexed territory included approximately 170,000 residents, were 77% white (compared with 62% citywide), and contained almost half of Atlanta's white residents. A similar pattern existed regarding income. The city's median family income stood at $5,029, yet 25 of the 33 census tracts had median incomes above the citywide level. Twenty-one exceeded the median family income for the metropolitan area, and the figure for 8 tracts was over $10,000. By 1970 the 46 tracts covering most of the land annexed in 1952 included over half of Atlanta's population. Moreover, 56% of the residents in the annexed area were white, compared with 49% for the city as a whole. Twenty tracts outstripped the metropolitan median family income level, including 7 greater than $15,000 (U.S. Bureau of the Census, 1962, 1972).

The lessons of the 1952 annexation were not lost on later politicians, who promoted territorial expansion for political or fiscal gain. Atlanta's last white mayor, Sam Massell, included in his 1973 reelection platform a call for annexation of a North Side area that would have added 50,000 whites to the city (Stone, 1989). Black leaders saw Massell's proposal as a way to maintain a white majority in the city. The tables had turned by the 1980s, when several black leaders actively discussed annexation of Sandy Springs, an affluent white area just north of the city, whose acquisition would add a substantial tax base but not threaten Atlanta's status as a black-majority city (Lorenz, 1989).

Demographic changes have been linked to public policies in other ways. Along with MARTA, city officials have promoted in-town housing for office workers around subway stops and on city-owned land near the civic center and other sites in the central business district (Saporta, 1990a). The city also has used mortgage revenue bonds to stimulate middle-class home ownership. Local officials also have tried to use such programs as public housing and urban renewal to aid black and poor white neighborhoods. Leaders have not hesitated, though, to tap such areas for commercial or industrial purposes when opportunities arose (Stone, 1989). Since Maynard Jackson's first term as mayor, city officials have attempted to use Hartsfield International Airport as a magnet for growth on the South Side and vigorously have opposed locating Atlanta's second airport near the affluent suburban areas north of the city.

Second, like its demographic changes, Atlanta's economic development has obvious ties to public policy. Business and political leaders have promoted Atlanta first as a regional, then a national, and, finally an international, city. During the Hartsfield and Allen mayoralties, industrial and infrastructure improvements received great attention. So did the formation of public-private partnerships and the growing involvement of state government in activities to improve Atlanta's economic fortunes (Henson & King, 1982). More recently, city and state officials have conducted highly visible trade missions to Europe and Asia. Former mayor Andrew Young capitalized on his previous position as United Nations ambassador to extend such efforts to Africa. Attempts to "internationalize" Georgia's economy with Atlanta as its gateway include the state's four overseas trade offices, one of them in Tokyo. Political and business leaders also have championed increased overseas service through Hartsfield International Airport, especially for locally headquartered Delta Air Lines.

Third, leaders seem to have concluded that the marriage of black political power and white-dominated investment is a positive sum game worth maintaining. As Stone (1989) has argued, this consensus and the use of selective benefits has placed the onus on the regime's challengers to dislodge it. Despite early resistance, the city's white business leaders have included blacks in their organizations and have made affirmative action an established municipal policy. This has allowed Atlanta's black mayors to improve minority representation in the municipal work force and to do more city business with minority and female firms (see Stein & Condrey, 1987). On the other side, while blacks appear to have a lock on the mayoralty and City Council, they are also supportive of a wide range of economic development policies.

REGIME CHANGE AND THE FUTURE OF ATLANTA POLITICS

There was an era when a white "power structure" could determine Atlanta's future through its control of private investment and City Hall. The long-term composition of Atlanta's regime shifted, though, when business leaders sought to use a coalition with the black middle class to ensure Atlanta's growth. That decision in the 1940s paved the way for the new regime that evolved during Maynard Jackson's tenure in the 1970s as growing black electoral power negated any chance of white control of City Hall. To the original governing coalition of the Chamber of Commerce, Central Atlanta Progress, the daily newspapers, major banks and retailers, and premier firms such as Coca-Cola have been added black politicians. Paradoxically, the shift of power within the city was followed closely by the growing influence of external actors—international firms, suburban officials, and the state government. Thus, while there is a new regime in Atlanta, its position for the near term will probably not mean "determining" the city's future so much as it will mean trying to steer Atlanta in a certain direction and then bargaining with a growing cast of decision makers outside the city.

NOTES

1. Although most of Atlanta's territory and residents are in Fulton County, portions of the city's East Side are in DeKalb County.

2. In discussing Atlanta, one must recognize the changing definition of the area. Data for the metropolitan area through the 1970s cover only 5 counties. With the 1980 Census, the Metropolitan Statistical Area (MSA) included 15 counties. Since 1983, the MSA has consisted of 18 counties. The "core counties" discussed below include the 7 (Clayton, Cobb, DeKalb, Douglas, Fulton, Gwinnett, and Rockdale) under the Atlanta Regional Commission, a regional planning agency.

REFERENCES

Allen, I., Jr., with Hemphill, P. (1971). *Mayor: Notes on the sixties.* New York: Simon & Schuster.

Atlanta Regional Commission. (1987). *Regional Transportation Plan, 1987-2010.* Atlanta: Author.

Bachtel, D. C. (Ed.). (1990). *The Georgia County Guide* (9th ed.). Athens, GA: Cooperative Extension Service, University of Georgia.

Banfield, E. C. (1965). *Big city politics.* New York: Random House.

Banks, M. E. (1987). *Consociational Democracy: The outcome of racial political polarization in Atlanta, Georgia, 1973-1986.* Unpublished doctoral dissertation, University of Texas at Austin.

Bayor, R. H. (1989). Urban renewal, public housing and the racial shaping of Atlanta. *Journal of Policy History, 1,* 419-439.

Bueno, J. (1990, May 28). Despite distance, Japanese bankers doing deals here. *Atlanta Business Chronicle,* pp. 1B, 3B-8B.

Bullock, C. S., III, & Campbell, B. A. (1984). Racist or racial voting in the 1981 Atlanta municipal elections. *Urban Affairs Quarterly, 20,* 149-164.

Clark, W.A.V. (1988). Racial transition in metropolitan suburbs: Evidence from Atlanta. *Urban Geography, 9,* 269-282.

Cline, K. (1987, April 13). Foreign banks boost local assets. *Atlanta Business Chronicle,* p. 3A.

Dedman, B. (1988a, May 1). Atlanta blacks losing in home loans scramble. *Atlanta Journal and Constitution,* pp. 1A, 14A

Dedman, B. (1988b, May 4). Bank protesters in Atlanta make ready to flex muscle. Atlanta Constitution, pp. 1A, 8A

Dedman, B. (1988c, May 2). Southside treated like banks' stepchild? *Atlanta Constitution,* pp. 1A, 4A-5A.

Dedman, B. (1988d, May 3). A test that few banks fail—in federal eyes. *Atlanta Constitution,* pp. 1A, 14A-15A.

Englade, K. (1985, October). Mixed reviews for Andrew Young. *Atlanta,* pp. 32, 37, 92-93, 123-125.

Feuerstein, A. (1990, July 16). Growing underclass Atlanta's Achilles' heel. *Atlanta Business Chronicle,* (Metro), p. 6.

The *Fortune* directory. (1965, July). *Fortune,* pp. 149-168.

The *Fortune* directory: Part II. (1965, August). *Fortune,* pp. 169-180.

The *Fortune* 500. (1990, April 23). *Fortune,* pp. 337-396.

The *Fortune* 500 largest service corporations. (1990, June 4). *Fortune,* pp. 304-332.

Georgia Secretary of State. (1989). *Georgia Official Directory of United States Congressmen, State and County Officers.* Atlanta: Author.

Georgia World Congress Center. (1989). *Annual Report 1989.* Atlanta: Author.

Hawkins, B. W., & Whelchel, C. (1968). Reapportionment and urban representation in legislative influence positions: The case of Georgia. *Urban Affairs Quarterly, 3,* 69-80.

Hayes, K. (1983, October 9). Campaign contributions: Where is the line? *Atlanta Journal and Constitution,* p. 14A

Henson, M. D., & King, J. (1982). The Atlanta public-private romance: An abrupt transformation. In R. S. Fosler & R. A. Berger (Eds.), *Public-Private Partnership in American Cities* (pp. 293-337). Lexington, MA: Lexington.

Hornsby, A., Jr. (1977). The Negro in Atlanta politics, 1961-1973. *Atlanta Historical Bulletin, 21,* 7-33.

Hutcheson, J. D., Jr., & Prather, J. E. (1988). Community mobilization and participation in the zoning process. *Urban Affairs Quarterly, 23,* 346-368.

Jones, M. H. (1978). Black political empowerment in Atlanta. *Annals of the American Academy of Political and Social Science, 439,* 90-117.

Key, V. O. (1949). *Southern politics in state and nation.* New York: Vintage.

Labich, K. (1989, October 23). The best cities for business. *Fortune,* pp. 56-93.

Lorenz, D. (1989, February 18). Sandy Springs incorporation effort: Democracy or racism? *Atlanta Constitution,* pp. 1A, 8A.

Murray, R., & Vedlitz, A. (1978). Racial voting patterns in the South: An analysis of major elections from 1960 to 1977 in five cities. *Annals of the American Academy of Political and Social Science, 439,* 29-39.

Pierannunzi, C., & Hutcheson, J. D., Jr. (1990). Electoral change and regime maintenance: Maynard Jackson's second time around. *PS: Political Science & Politics, 23,* 151-153.

Rice, B. R. (1981). The battle of Buckhead: The Plan of Improvement and Atlanta's last big annexation. *Atlanta Historical Journal, 25,* 5-22.

Sack, K. & Copeland, L. (1989, April 23). Incumbency provides big boost to Lomax's campaign financing. *Atlanta Journal and Constitution,* pp. 1A, 6A.

Saporta, M. (1989, January 9). The myriad business voices. *Atlanta Constitution,* pp. B1, B8.

Saporta, M. (1990a, May 20). Intown housing market no longer on the outside. *Atlanta Journal and Constitution,* pp. H1, H3.

Saporta, M. (1990b, July 11). Japanese investment in Georgia slows. *Atlanta Constitution,* p. A9.

Saporta, M. (1990c, January 29). New leader wants to get CAP moving and shaking again. *Atlanta Constitution,* pp. A1, A11.

Schwartzkopff, F., & Yancey, W. R. (1990, October 12). MARTA's biggest foe may be racism. *Atlanta Constitution,* pp. C1, C3.

Sherman, M. (1990, April 23). Ethics of council donations questioned. *Atlanta Journal and Constitution,* pp. A1, A13.

Stein, L., & Condrey, S. E. (1987). Integrating municipal workforces: A comparative analysis of six Southern cities. *Publius, 17,* 93-103.

Stone, C. N. (1976). *Economic growth and neighborhood discontent: System bias in the urban renewal program in Atlanta.* Chapel Hill: University of North Carolina Press.

Stone, C. N. (1989). *Regime politics: Governing Atlanta, 1946-1988.* Lawrence, KS: University Press of Kansas.

Thomas, E., Jr. (1989, October 2). Connections foster deals. *Atlanta Business Chronicle,* pp. 1A, 24A-36A.

Thomas, E., Jr. (1990, July 23). New cash options make city council perk up. *Atlanta Business Chronicle,* p. 6A.

U.S. Bureau of the Census. (1962). *Census tracts: Atlanta, Ga. Standard metropolitan statistical area. U.S. censuses of population and housing: 1960* (Final Report PHC[1]-8). Washington, DC: Government Printing Office.

U.S. Bureau of the Census. (1963). *Census of population: 1960. Vol. 1: Characteristics of the population, Part 12: Georgia.* Washington, DC: Government Printing Office.

U.S. Bureau of the Census. (1964). *Census of population: 1960. Vol. 1: Characteristics of the population, Part 1: United States summary.* Washington, DC: Government Printing Office.

U.S. Bureau of the Census. (1968). *1967 census of governments. Vol. 2: Taxable property values.* Washington, DC: Government Printing Office.

U.S. Bureau of the Census. (1970). *1967 census of governments. Vol. 7: State reports, No. 10: Georgia.* Washington, DC: Government Printing Office.

114   BIG CITY POLITICS

U.S. Bureau of the Census. (1971). *Local government finances in selected metropolitan areas and large counties: 1969-70.* Washington, DC: Government Printing Office.

U.S. Bureau of the Census. (1972). *Census tracts: Atlanta, GA, Standard metropolitan statistical area. 1970 census of population and housing* (Report PHC[1]-14). Washington, DC: Government Printing Office.

U.S. Bureau of the Census. (1983). *1980 census of population. Vol. 1: Characteristics of the population, Chap. C: General social and economic characteristics, Part 12: Georgia.* Washington, DC: Government Printing Office.

U.S. Bureau of the Census. (1986). *State and metropolitan area data book, 1986.* Washington, DC: Government Printing Office.

U.S. Bureau of the Census. (1989a). *1987 census of governments. Vol. 2: Taxable property values.* Washington, DC: Government Printing Office.

U.S. Bureau of the Census. (1989b). *City government finances in 1986-87.* Washington, DC: Government Printing Office.

U.S. Bureau of the Census. (1990a). *1987 census of governments. Vol. 4: Government finances, No. 3: Finances of county governments.* Washington, DC: Government Printing Office.

U.S. Bureau of the Census. (1990b). *1987 census of governments. Vol. 4: Government finances, No. 5: Compendium of government finances.* Washington, DC: Government Printing Office.

U.S. Bureau of the Census. (1990c). *Local population estimates. South: 1988 population and 1987 per capita income estimates for counties and incorporated places* (Series P-26, No. 88-S-SC). Washington, DC: Government Printing Office.

U.S. Bureau of the Census. (1990d). *Local population estimates. Population estimates for metropolitan statistical areas, July 1, 1988, 1987, and 1986* (Series P-26, No. 88-B). Washington, DC: Government Printing Office.

Walker, T. (1988, November 13). Atlanta's civic power now held by many hands. *Atlanta Journal and Constitution,* pp. 1E, 5E.

8

Miami: Minority Empowerment and Regime Change

RONALD K. VOGEL
GENIE N. L. STOWERS

MIAMI HAS UNDERGONE a radical transformation in the last three decades. No aspect of its political, economic, or social condition has been left unchanged. The most striking difference is to be found in Miami's ethnic composition. Thirty years ago, Hispanics were an insignificant proportion of the population. Today, Cubans are a majority within the city of Miami and have come to dominate its government institutions. By the year 2000, they will comprise a majority of the county population as well.

The Cubans and other Latins have had a dramatic effect upon the fortunes of downtown and have been a major factor in redefining Miami as an international city. Miami is not just bilingual, it is trilingual, with Creole as the newest addition to the area's English and Spanish languages. The immigration of peoples from the Caribbean basin and other Latin and Central American countries has resulted in the creation of a multiethnic community that serves as the "window to Latin America." The broadening of Miami's tourism base, as well as the rise of international banking, have given Miami a new role in the international division of labor. This chapter highlights the political, social, and economic changes that Miami experienced during this transition.[1]

AUTHORS' NOTE: The authors gratefully acknowledge research support they received for this study. Stowers received a grant from the American Political Science Association, 1989. Vogel received research support from the University of Louisville College of Arts and Sciences Dean's Professional Research Initiative, Spring, 1990, and the University of Louisville President's Commission on Academic Excellence Summer Fellowship, 1990. The order of the authors' names was determined by the toss of a coin.

DEMOGRAPHIC AND SOCIAL CHANGE

THE DEVELOPMENT OF A MULTIETHNIC CITY

Several characteristics distinguish the Miami of today from other American cities and highlight the dramatic changes of the past 30 years. Miami differs from most of the other metropolitan areas in this study in that the city makes up only a small part, about 20%, of the county.[2] The Miami metropolitan area has no majority population group. Of those U.S. cities with a Spanish-origin population of 150,000 or more, only San Antonio and El Paso have higher proportions of Hispanics in the central city (McCoy & Gonzalez, 1985).

Immigration changed Miami from a relatively homogeneous, small, Southern town dominated by non-Latin whites to a multiethnic, trilingual (English, Spanish, and Creole) community. Between 1960 and 1990, Dade County doubled its population, rising to 1.912 million people. Most of this population growth can be attributed to Hispanic, primarily Cuban, immigration (Metro-Dade County Planning Department, 1985).

It is estimated that anywhere from 600,000 to 775,000 Cuban immigrants came to Miami from 1962 through 1980. The newest Latin immigrants are the Columbians and Nicaraguans, who began arriving in the 1980s. Figure 8.1 and Table 8.1 show the extent of ethnic and racial changes experienced in the Miami area between 1960 and 1990.

In this same period, the county's black population tripled through natural increase and the arrival of as many as 80,000 Haitians (many undocumented) in South Florida (Stepick & Portes, 1986). The county's non-Latin white population ("Anglos") declined between 1960 and 1990 (see Figure 8.1). White flight is certainly occurring. Internal Revenue Service data indicate 45,000 persons moved from Dade County to four other Florida counties between 1980 and 1985. Voter registration data reveal that 84% of the voters moving out of Dade County in 1984 were non-Hispanic whites, providing even greater evidence of white flight (Dugger, 1987).

About one third of Dade's non-Latin white population is Jewish. Although Jews still make up about 13% of Dade County's population, their proportion of the population has declined significantly because of deaths and changes in Jewish migration flows (Sheskin, 1990).

SOCIAL STRAIN

The need to assimilate an estimated 679,000 new immigrants with different cultures and languages into one community in one 30-year period has resulted in serious social strain. The impact of these refugees on Miami's city

Race & Ethnicity
City of Miami

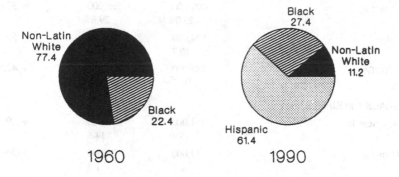

1960   1990

Race & Ethnicity
Dade County

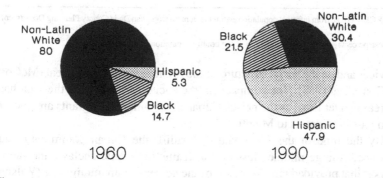

1960   1990

Figure 8.1.

NOTE: Dade County figures, 1960-1980, are from Metro-Dade County Planning Department, Research Division, *Dade County Facts*, 1990; 1990 figures are preliminary U.S. Census estimates from Metro-Dade County Planning Department. Miami figures are from the City of Miami Planning Department—U.S. Census data (1960-1980) and their own estimates (1990).

TABLE 8.1

Diversity Among Dade County Racial and Ethnic Groups, 1980 and 1989 (Estimated)

	1980	1989	% change
HISPANIC GROUPS:			
Cuban	407,000	560,000	+ 37.6
	25.0%[1]	29.8%	
Other	174,000	270,000	+ 55.2
	10.7	14.4	
TOTAL	581,000	830,000	+ 42.9
	35.7	44.1	
BLACK GROUPS:			
American Black	251,000	274,000	+ 9.2
	15.4	14.6	
Hispanic	11,000	26,000	+136.4
	0.7	1.4	
Haitian	18,000	80,000	+344.4
	1.1	4.3	
TOTAL	280,000	380,000	+ 35.7
	17.2	20.2	
NON-LATIN WHITE GROUPS:			
TOTAL	765,000	670,000	– 12.4
	47.0	35.6	
TOTAL POPULATION	1,626,000	1,880,000	+ 15.6
	99.9	99.9	

SOURCE: Derived from informal population estimates from Metropolitan Dade County Planning Department, 1989.

[1] These percentages reflect the proportion of the county population of these groups.

services and on the social structure of the Miami area has been great (McCoy & Gonzalez, 1985). Their impact on the social well-being of the area has increased over time, as the newer Cuban and Haitian immigrants are poorer than past immigrants to Miami.

By the time of the 1980 Mariel boatlift, the Cuban community had produced a large middle class and a distinctive ethnic enclave and labor market that provided jobs for many of the newer Cuban immigrants (Wilson & Portes, 1980). Cuban refugees were provided with jobs and quickly integrated into the economic community through already-established businesspeople. Cuban businesses also benefited by gaining a continuing supply of cheap labor.

Still, more difficulties were experienced in integrating the Mariel refugees in 1980 than with any other group of Cubans. They tended to be younger, more likely to be without family members in the United States, were the first wave of immigration to include sizable numbers of blacks, and had the added misfortune of having lived under Castro's Cuba—a fact that many in Miami found difficult to accept.

Cuban refugees in general received federal welfare assistance, job training and education subsidies, and funds to help meet professional-certification requirements during their resettlement. This assistance, $957.1 million from 1961 until 1974, provided an important lift for the Cuban community's eventual economic growth and success of its businesses (Pedraza-Bailey, 1985). This experience is in sharp contrast to the 1.2% of Haitian refugees who are employed by other Haitian refugees. According to a 1986 study, 63% of Haitian refugees were jobless and 59% were below the poverty line (Stepick & Portes, 1986).

American blacks have not fared as well as the Cubans over the last three decades. They resent the retraining and resettlement funds obtained by Cuban immigrants and their rapid attainment of economic and political power. The black community also perceives the Cubans as having gained at the expense of the black community. They believe the Cuban community is unwilling to hire blacks and that the prevalence of bilingualism precludes their obtaining jobs even in the service industries.

The so-called "Cuban success story" reminds the black community of how little it has progressed in the last 30 years, in spite of progress by blacks in other areas of the country. The Cubans began to arrive in the United States only five years after the 1954 *Brown v. Board of Education* decision. In the rest of the country, black communities in the 1960s were beginning to gain political power. But in Miami, blacks watched in frustration as Cubans passed them by. Early (but brief) biracial coalitions gave way to a dominant Cuban majority, politically and socially.

The different rates of group economic development also resulted in sharp differences between the groups in income and poverty levels. According to the 1980 Census, the median family income in 1980 for blacks was only $13,108. Median family income for Hispanics was $15,749, 20% higher than the level for blacks. For non-Latin whites, median family income was $19,585, 50% greater than the level for black families (U.S. Bureau of the Census, 1982).

These inequalities and perceptions of group differences led to high levels of resentment in the community. This resentment has been manifested most explicitly in the riots in the black community over the past 20 years. Porter and Dunn (1984) report at least 13 violent episodes between blacks and whites from 1970 to 1979. Another series of riots occurred between 1980 and 1989, the most recent being the week before the Super Bowl in 1989. The 1989 riot may have been related, in part, to anger in the black community over federal aid provided to new Nicaraguan immigrants. The riots in the 1980s were triggered by specific events, often involving the shooting of blacks by non-Latin white or Hispanic police officers. Underlying factors, however, were at least as critical as any precipitating event in creating this violent behavior in Miami. Porter and Dunn believe that the lack of black access to the system in Miami and the displacement of blacks by Cubans from jobs and other opportunities were the underlying factors behind at least the 1980 riot.[3]

The U.S. Commission on Civil Rights (1982) warned that unless steps were taken to address these problems, renewed violence was likely. Although a number of governmental and private-sector programs were put into place in response to the Commission's report, the commitment of leaders, resources, and follow-through has been too uneven to have much effect (Bivens, 1983; Dugger, 1985).

Over the past 30 years, much resentment and discomfort between blacks, Hispanics, and non-Latin whites have developed. Blacks resented being left out of both the political and economic success. Hispanics felt they were not given enough credit for their part in creating the Miami success story, particularly the development of its new image as the "international city" and bridge to Latin America. Non-Latin whites who had formerly controlled the city economically and politically were faced with losing control of their city. Many believed that the newcomers should have assimilated into the Miami community. They felt that they and Miami had been forced to change into a community where speaking only English could mean not getting a job and not being able to access services.

The 1980 antibilingual ordinance, initiated by a petition drive and passed in a county referendum, illustrates the depth of resentment toward Latins in the community (Santiago, 1983). The ordinance was intended to limit county expenditures for any multicultural or multilingual purposes. Although it actually has had little effect in the county, apparently not all residents of Dade County favor the establishment of an "international" city. Miami has faced many changes in its social fabric over the past 30 years and is struggling to find ways to adjust to them. These social changes have been mirrored by equally startling political changes.

POLITICAL CHANGE

FORMS OF GOVERNMENT

Both Miami and Dade County are products of the reform movement. Both have council-manager forms of government, at-large elections, and nonpartisan races. In the city of Miami are four commissioners and a mayor, all elected at-large. All eight county commissioners and the elected county mayor are also elected at-large. Neither mayor has administrative powers. Ironically, the mayor of the city of Miami has greater visibility than the mayor of the larger Metro-Dade County.

The position of manager in each government is very important. An unforeseen consequence of the council-manager form of government, followed by both governments, is that the appointments to the posts of city and county manager have become important avenues of political incorporation for minorities and figure prominently in electoral coalition building and regime structuring (see Stone, 1989).

THE METROPOLITAN EXPERIMENT

Dade county is cited frequently as an "experiment" in two-tiered metropolitan government (see Sofen, 1966; Lotz, 1984). The "home-rule" charter adopted in 1958 established the framework for a two-tier system of service provision with local government (city) providing the "life-style" functions and Metro government (county) providing the "system-maintenance" functions (Harrigan, 1989). The Metro charter does not delineate clearly which services are areawide and which are local, and controversy still exists over how to differentiate these services. In the early years of Metro, a number of court cases were required to establish Metro's preeminence. Even today, Metro is viewed with some ambivalence by city officials.

Gradually, Metro has acquired responsibility for additional services, some of which local governments voluntarily relinquished—traffic engineering, water and sewer, solid waste disposal, housing, libraries, some aspects of law enforcement (e.g., crime lab), and transportation. Fire protection is the best example of a service that small municipalities have been eager to see Metro take on (Lotz, 1984).

Although the "metropolitan experiment" may have been an improvement over the previous political arrangements, Metro-Dade government is not truly a "metropolitan" government, and the "two-tier" system is not in place for almost half the residents of Dade County. Changing growth and settlement patterns and the refusal of Metro-Dade government to incorporate new cities has resulted in half the residents of the county—those in unincorporated areas—having only one level of government (the county) providing them with services.

The white business and civic elite dominated Miami politics from Metro's inception (Warren, Corbett, & Stack, 1990). Metropolitan government, although spearheaded by traditional reform groups (the League of Women Voters, university professors, civic and professional associations, the newspapers), served the business community's interest in centralized government that could provide the public services and the physical infrastructure needed because of rapid growth (Sofen, 1966; Lotz, 1984).

There is no doubt that Metro government works against minority interests (see Mohl, 1989; Warren et al., 1990). Blacks made efforts to incorporate Liberty City, following the riots in the early 1980s. Their complaints included the quality of housing, police brutality, and poor transportation and other services in this area (Mohl, 1989). In fact, an attempt was made by blacks in the late 1980s to incorporate Liberty City to gain more autonomy.

THE NEW REFORM MOVEMENT

A 1984 University of Miami study, commissioned by the business community, recommended the adoption of a strong-mayor system of government and single-member districts[4] and some merger—the locus of a "new reform" movement. The county commission refused to place this proposal on the ballot. Mohl concluded that this plan "most likely would put the old Anglo power structure back in control" (Mohl, 1989, pp. 150-151).

Ironically, it is the *Miami Herald* and the business community, the strongest advocates of "good government" associated with the earlier reform movement, that are now so supportive of the strong-mayor system and single-member district elections (Mohl, 1989). Initial downtown business opposition to single-member districts stemmed from its ability to control the Metro-Dade Commission. Single-member districts, however, have never been approved by the voters. After several years in the courts, the Metro-Dade Commission has been forced to pursue a settlement of a Voting Rights Act case (Meek et al. v. Metro-Dade County) filed by black and Hispanic leaders against the county. As a result, the county will have at least some single-member election districts to ensure election of blacks and Latins to the Metro-Dade Commission.

THE ETHNIC BASE OF THE VOTE

Electoral politics also have been affected by the changing social composition of the community. The changing ethnic and racial balance began to alter the political balance of power on the Miami City Commission in the early 1960s. Figure 8.2 shows the changes in voter registration in the city of Miami from 1975 until 1989. Baseline data indicate that in 1956, whites in the county constituted 94.2% of registered voters; blacks were the other 5.8%.[5] By 1975, blacks had registered in far greater numbers. By this time, Cubans also began to make their presence felt as they become naturalized citizens and registered to vote.

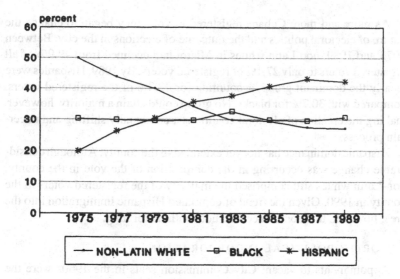

Figure 8.2. Voter Registration by Race and Ethnicity, Miami, 1975-1989. Derived from Metropolitan Dade County Election Department Data.

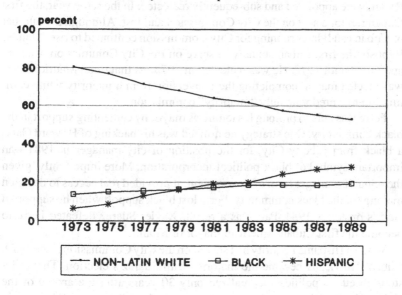

Figure 8.3. Voter Registration by Race and Ethnicity, Dade County, 1973-1989.

As more and more Cubans registered to vote, they began to change the nature of electoral politics and the outcome of elections in the city. Between 1973 and 1989, non-Latin whites in Miami had declined from 49.9% of all registered voters to only 27.1% of registered voters. By 1989, Hispanics were clearly the dominant group, accounting for 42.3% of the registered voters, compared with 30.7% for blacks. No group could attain a majority, however, making the creation of electoral coalitions a continually shifting and uncertain process.

Hispanic dominance has not yet extended to the county. Although considerable change was occurring in the composition of the vote in the county, non-Latin whites still comprised the majority of the registered voters in the county in 1990. Given the trend of continued Hispanic immigration into the area, however, this dominance will not continue for long.

MINORITY POLITICAL INCORPORATION

Appointments to vacant City Commission seats in the 1960s were the major factor in political incorporation of minorities in Miami. In 1967, the first black, Athalie Range, and the first Hispanic, Maurice Ferre, a Puerto Rican, were appointed and subsequently reelected. In the same year, the first Cuban ran for a seat on the City Commission and lost. Although the number of Cuban candidates running for City Commission continued to rise, Manolo Reboso, the first Cuban actually to serve on the City Commission, was not appointed until 1972. He was reelected in 1973. At that time, Maurice Ferre was elected mayor, completing the conversion from a majority white commission to a predominantly "minority" commission.

Ferre was able to prolong his tenure as mayor by cementing support in the black community. One strategy he utilized was his backing of Howard Gary, a black from Liberty City, for the position of city manager in 1981—an important symbol of black political incorporation. More importantly, given the council-manager form of government, it provided real access to decision making for the black community. Ferre lost black support when he supported Gary's ouster in 1984. Partly as a result, Xavier Suarez defeated Ferre to become the first Cuban-American mayor of Miami.

While a Hispanic majority had served on the City Commission since 1979, Cuban-Americans became the majority with Suarez's election. This milestone in ethnic politics was realized only 30 years after the arrival of the Cubans in America, making them the most politically successful group in American history. One black and one non-Latin white also have continued to retain a seat on the Commission. Since these elections are at-large and voters from these two groups are so outnumbered, the continued presence of representatives from these current minority groups is a curiosity.

The Cuban community, secure in its electoral position in Miami, already has begun to separate along generational, ideological, class, and other lines. This is most readily apparent in the growing number of campaigns pitting Cuban-Americans against each other. However secure the Cuban-American position in Miami, race and ethnicity will continue to be the dominant cleavage there. In a city with no electoral majority and changing racial and ethnic registration patterns due to continued immigration, the electoral coalition needed by candidates for at-large seats change constantly. No candidate since the mid-1970s has been able to put together a majority coalition with just one group. But it is not uncommon for electoral support among the three groups to shift from one election to another, including runoff elections.

IMPACT OF THE CHANGING ETHNIC BASE UPON MIAMI'S POLITICS

The Cubans have changed the character of Miami's city elections by running on foreign-policy platforms instead of local issues. From 1973 until the present, elections have been characterized by increased but varying levels of ethnic and racial hostility and tension and have been heavily ideological. Candidates could improve their standing in the Cuban community and win elections by tapping into the anti-Castro feeling and publicizing their counterrevolutionary credentials.

In recent years, successful Cuban candidates have focused upon more traditional local issues, suggesting assimilation into the American political system. This shift also is associated with the new generation of young Cuban-Americans now running for office. They were born or raised in the United States and have little personal experience with Castro and the revolution. These younger Cuban-Americans are often more familiar with the American political experience than with events in Cuba more than 30 years ago (Stowers, 1988; see also Mohl, 1989).

Such emotional issues as Castro or communism have had their impact upon elections in Miami in other ways. Unlike in many other American cities, voter turnout in Miami is extremely high, suggesting a high degree of ethnic conflict and intergroup competition. Since 1973, voter turnout among Cuban-Americans has exceeded that of any other group in Miami's elections, reaching 69% in the 1983 mayoral runoff election. Black-Americans also experienced high levels of turnout, equaling or surpassing non-Latin whites in most elections from 1981 until the present (Stowers, 1990).

The entrance of Cuban-Americans into the political arena also has altered the partisan alignments of Miami and Dade County. Like the rest of the South, Dade County's partisan orientation had been overwhelmingly Democratic.

In 1956, the county was 87% Democratic and 13% Republican. But by 1990, the county was 55.5% Democratic and 36.8% Republican. The growing strength and involvement of the conservative Cuban-American community has had important consequences for major party representation in the state legislature and even the Congress.

These shifts have had the effect of further increasing the social and political distances between various groups, as political party affiliation has reinforced ethnic cleavages, rather than reducing them. Much of this shift was caused by the increased registration of conservative Cuban-Americans as Republicans. In 1990, Hispanics registered 68.8% Republican. Blacks were still overwhelmingly Democratic, 92.7%, while non-Latin white Democratic registration was 60.3%. Black Democratic registration has declined slightly since 1983. Non-Latin white Democratic registration declined 10% in this same time period. Hispanics, with their growing numbers and heavy turnout patterns, increased from 50.7% to 68.8% Republican. Thus, the Republican Party has gained as Cuban-Americans have registered Republican and become more active in party politics.

Because of these drastic shifts in partisanship and the increased participation of Cuban-American voters, Cuban-Americans have gained many seats in the state legislature, elected on a Republican label. The impact of the Cuban vote upon Democratic fortunes was highlighted by Republican Cuban-American Ileana Ros-Lehtinen's 1989 capture of Claude Pepper's seat in Congress.

POLITICS IN THE COUNTY

The political realities of voter registration, electoral coalitions, and ethnic and racial representation in the city are in sharp contrast to those at the county level. Even today, there is only one Hispanic and one black member on the County Commission. Each of these was appointed initially to the seat and has since been reelected, thus continuing the pattern of minority political incorporation through appointment to vacant seats. In the short run, the prospect of single-member districts should enhance black and Hispanic political power in the county. But the likelihood of further Latin immigration suggests Hispanics ultimately will be the dominant group in the county, as well as in the city.

Hispanics already are capitalizing on their economic successes to influence County Commission races through campaign finance. The fact that the Latin Builders Association was able to get the County Commission to extend urban boundaries into previously reserved areas in the county is one indication of the success of this strategy. It also points out how economic power can be translated into political power. The black community has not been able to replicate this approach.

The consequences of the differing voter registration and representation patterns in the city and the county can be seen also in some public-policy outcomes. Historically, public sector jobs have been an important source of upward mobility for minority groups. In the 1970s, both governments were dominated by non-Latin white employees, while blacks had approximately 30% of the jobs. By 1990, non-Latin whites had lost a significant share of jobs (from 50.2% to 29.8%), and Hispanics had gained significantly in local government employment.

Significant differences developed, however, between the two governments in the levels of Hispanics and non-Latin whites employed. While 38.8% of city employees were Hispanic in 1990, the county employed only 29.7% Hispanic. The level of black employment has remained stable at approximately 30%. These differences indicate some higher response to Hispanics by the city, with a political body dominated by Cuban-Americans.

THE LOCAL REGIME

Stone defines a regime as "the informal arrangements by which public bodies and private interests function together in order to be able to make and carry out governing decisions" (Stone, 1989, p. 6). The main task of a regime—governing—involves "managing conflict" and "making adaptive responses to social change" (Stone, 1989, p. 6). Community direction may be provided by the private power structure, local government, or both cooperatively.

Sometimes a regime structure is unable to form or to maintain itself, leaving a community essentially ungoverned, a situation akin to "hyperpluralism" (see Vogel, forthcoming). This is the position Miami finds itself in today. As Koenig points out, "A fading power structure and weak government have created a leadership vacuum in Florida's most volatile metro area" (Koenig, 1990, p. 42). Before the arrival of the Cubans and Miami's dramatic changes, the city had a power structure that effectively "managed conflict" and saw that the community "adapted" to changes in the larger society and economy. This is no longer the case, however.

The power structure was organized as the "Non-Group," established in 1970 under the direction of Alvah Chapman, chairman of Knight Ridder Newspapers. As one member of the Non-Group stated, "We could move a mountain pretty fast when there were 12 of us. . . . The civic agenda was dictated by the power structure. The government put it on the ballot. And the *Herald* said, 'Vote for it.' That won't work anymore" (Duggar, 1988). A *Miami Herald* study points out that "no individual or group can stake a convincing claim to civic leadership of Dade's fragmented community" (Dickerson, 1988).

The influence of the power structure has declined partly because of the withdrawal of some of the more influential business leaders through retirement (e.g., Alvah Chapman of the *Miami Herald*) and the changing fortunes of some of Miami's dominant businesses (e.g., Eastern Airlines). The business elite are still organized and essentially cohesive but now lack the resources to mobilize an increasingly heterogeneous community behind its agenda. This has less to do with changes in the nature of the existing power structure than with changes in the broader community in the 1980s.

Thompson (1984) identifies the groups that the power structure was best able to influence: the Miami City Commission, the Metro-Dade County Commission, the Florida legislature, business, and the media. Its influence is much weaker over citizen voters, particularly blacks and Hispanics. As voters, particularly the Latin voters, have become more dominant in city elections, the power structure has found its power declining in the city of Miami, and later in the county.

Today, ethnic and racial group identification and diversity are the predominant features in the area's regime politics. The decline of the *Miami Herald* as the major source of news for much of the population and the creation of alternative sources of power within the Latin community, including several influential radio stations and newspapers, have undermined seriously the capacity of the traditional power structure to govern.

The Non-Group did not recognize its declining influence until the failure of the occupational tax that it pushed to finance an orchestra. At that point, the Non-Group realized it did not have contacts in the Cuban community. In an effort to incorporate new power centers, the Non-Group enlarged its membership. But the power structure may have waited too long. The Cuban community already had developed alternatives to the Non-Group in the Latin Builders Association and the Cuban American National Foundation, both of which are increasingly knowledgeable politically.

The Cubans dominate city government and are in a position to challenge non-Latin white power in county government. But they have not challenged the conservative values and policies of the community and have shown a propensity to form coalitions with the white business community (Warren et al., 1990). The Cubans, then, dominate the public side of the emerging regime structure. The private power structure still holds financial and economic power that public officials and other community groups must access in order to accomplish their own goals.

A major difficulty in forging a new regime structure is the inability of the mayor of Miami to negotiate an enduring accommodation with the power structure because of the volatility of electoral coalitions. This inability probably underlies the power structure's support for a strong-mayor system for the two governments in order to create a stronger leadership position in the more pluralistic and complex environment that Miami now faces. The fact that ethnic cleavages at the mass level carry over to the elites is a further complication in forging a durable regime structure. What is most likely to occur is a period of change and fluctuating coalitions with the continuing decline in the influence of the earlier non-Latin white power structure as Hispanic economic power and political influence grow.

CONCLUSION

The real challenge for Miami and Dade County is to more fully incorporate blacks into the economic and political system. Cubans gained their political power after establishing a firm economic base. But the Cuban example is not particularly relevant to the black community, which lacks the professional skills and community ties that the Cubans brought with them.

Greater political incorporation that should have followed civil rights gains in the 1960s was blocked by the sheer size and economic successes of the immigrant population. Blacks have neither economic nor political power. If the political system does not become more responsive to black concerns, it is likely that the racial disturbances and riots of the 1980s will continue into the next century. But these violent responses have not and probably will not produce the kind of sustained attention and policies necessary to bring about black political incorporation.

Hispanics do not have a great deal of incentive to bring blacks into the political system. The downtown white business structure may have reached an accommodation with the Cuban community that does not threaten its own interests, but this accommodation does leave out the black community. The community's growth strategy is based firmly on pursuing international trade and taking advantage of its proximity to Latin and Central America.

Events on the horizon have the potential to change drastically at any time the political situation in Miami. The continued immigration of groups with varying economic and political bases or the much-awaited fall of Castro could once again alter the political, economic, and social fabric of Miami in the future. But it is hard to imagine more drastic change than Miami has experienced in the last 30 years.

NOTES

1. This study was completed utilizing multiple methods of inquiry, including primary data collection, examination of secondary and historical sources, and interviews with political, economic, and civic leaders, as well as leaders in the black and Latin communities. Stowers conducted 25 interviews in Miami in late 1988 and early 1989. Vogel interviewed 16 additional persons in the spring and early summer of 1990.

2. The Miami MSA includes Monroe County as well as Dade County. South Florida usually is considered to be Dade, Broward, and Palm Beach Counties.

3. Porter and Dunn (1984) found that the 1980 riots were unusual in that whites were the target of the violence and the rioters were from a more law-abiding group than those who participated in earlier riots.

4. Single-member districts for Metro-Dade government were rejected in a countywide referendum in 1971. Single-member districts were considered again by a charter review commission. Mayor Ferre had advocated single-member districts in 1985 (Mohl, 1989). County voters again rejected single-member districts in 1990 after yet another charter review commission.

5. Although the figures are not available, it is highly probable that the proportions were similar for the city of Miami.

REFERENCES

Bivens, L. (1983, November 27). Promises and plans: Where they stand. *Miami Herald*, p. 1A.

Dickerson, B. (1988, January 31). Miami's power elite. *Miami Herald*, p. 1A.

Dugger, C. (1985, May 13). Lofty plans helping few Dade blacks. *Miami Herald*, p. 1A.

Dugger, C. (1987, May 31). Latin influx, crime promote 'flight' north. *Miami Herald*, pp. B1, B2.

Duggar, C. (1988, February 1). New rivals challenge the old guard. *Miami Herald*, p. 1A.

Harrigan, J. (1989). *Political change in the metropolis* (4th ed.). Glenview, IL: Scott, Foresman.

Koenig, J. (1990, February). Dade County: Who's in charge here? *Florida Trend*, pp. 42-46.

Lotz, A. (1984). *Metropolitan Dade County: Two-tier government in action.* Boston: Allyn & Bacon.

McCoy, C. B., & Gonzalez, D. H. (1985). *Cuban immigration and immigrants in Florida and the United States: Implications for immigration policy* Gainesville, FL: University of Florida Bureau for Economic and Business Research.

Meek et al. v. Metro-Dade County, Florida, et al., 86-1820-CIV-Ryskamp.

Metro-Dade County Planning Department, Research Division. (1985). *Demographic profile, Dade County, Florida: 1960-1980.* Unpublished mimeograph.

Metro-Dade County Planning Department, Research Division. (1989, January). *Unofficial population estimates.*

Metro-Dade County Planning Department, Research Division. (1990). *Dade County facts.* Unpublished mimeograph.

Mohl, R. (1989). Ethnic politics in Miami, 1960-1986. In R. M. Miller & G. E. Pozzetta, (Eds.), *Shades of the Sunbelt* (pp. 143-160). Boca Raton: Florida Atlantic University Press.

Pedraza-Bailey, S. (1985). *Political and economic migrants in America: Cubans and Mexicans.* Austin: University of Texas Press.

Porter, B., & Dunn, M. (1984). *The Miami riot of 1980.* Lexington, MA: Lexington.

Santiago, F. (1983, May 4). Blacks feel overwhelmed by Cubans, leaders say. *Miami Herald*, p. 1A.

Sheskin, I. (1990). The Jews of south Florida. In M. Castro (Ed.), *Documenting Dade's diversity: An ethnic audit* (pp. 1-8). Miami: Greater Miami United.

Sofen, E. (1966). *The Miami metropolitan experiment*. Garden City, NY: Anchor Books.

Stepick, A., & Portes, A. (1986). Flight into despair: A profile of recent Haitian refugees in south Florida. *International Migration Review, 20*(2), 329-350.

Stone, C. (1989). *Regime politics: Governing Atlanta*. Lawrence, KS: University Press of Kansas.

Stowers, G.N.L. (1988, September). Ethnic and racial candidates in the transition to urban political power. Paper presented at the American Political Science Association Annual Convention, Washington, DC.

Stowers, G.N.L. (1990). Cuban political participation and class status. *Ethnic Groups, 8*(2), 73-90.

Thompson, S. A. (1984). Community leadership in greater Miami, Florida: What role for blacks and Cuban Americans? Unpublished doctoral dissertation, Southern Illinois University.

U.S. Bureau of the Census. (1976a). *Census of Population, 1960: Florida*. New York: Arno Press.

U.S. Bureau of the Census. (1976b). *1970 Census of Population: Florida*. New York: Arno Press.

U.S. Bureau of the Census. (1982). *1980 Census of Population: Florida*. New York: Arno Press.

U.S. Commission on Civil Rights. (1982). *Confronting racial isolation in Miami*. Washington, DC: Author.

Vogel, R. K. (forthcoming). *Local political economy*. Gainesville, FL: University of Florida Press.

Warren, C., Corbett, J. G., & Stack, J. F., Jr. (1990). Hispanic ascendancy and tripartite politics in Miami. In R. P. Browning, D. R. Marshall, & D. Tabb (Eds.), *Racial politics in American cities* (pp. 155-178). New York: Longman.

Wilson, K., & Portes, A. (1980). Immigrant enclaves: An analysis of the labor market experiences of Cubans in Miami. *American Journal of Sociology, 86*, 295-319.

9

New Orleans: The Ambivalent City

ROBERT K. WHELAN
ALMA H. YOUNG

DEMOGRAPHIC AND INDUSTRIAL UPHEAVAL

DEMOGRAPHIC CHANGE

New Orleans's demographic trends of the last 30 years resemble those of the declining industrial cities of the Northeast and the Midwest in a number of respects (Mumphrey & Moomau 1984). In the decade 1970-1980, the city's population decreased from 593,471 to 557,616, a decline of 6.1%. The 1990 figures show a population of 497,000, a decline of 21% from the previous decade. The city is part of a Metropolitan Statistical Area (MSA) of 1,324,400 persons, currently making it the 35th-largest metropolitan area in the United States. Although the suburban areas of the MSA evidenced slow population growth in the 1980s, this growth was concentrated in the early part of the decade, and that growth was slower than in other mid-South metropolitan regions.

The city population pattern in the decades under review has been one of decline and stagnation. In the 1970s this population change was essentially white flight to the suburbs, especially Jefferson Parish, which is directly west of New Orleans (Orleans Parish). Most recently, in the 1986-1988 period, the New Orleans MSA suffered an average annual net outmigration of about 8,000 people, as they sought jobs in other regions.

New Orleans became a black-majority city in the 1970-1980 decade, changing from 45% black and 54% white in 1970, to 55% black and 43% white in 1980. In the most recent census estimates, minorities comprise 61% of the population in New Orleans. Registered voters have reflected the same trends. New Orleans has had a tradition of black voter registration since the 1940s. In 1960 blacks comprised 16.9% of the city's registered voters. By the 1986 mayoral election, blacks were a majority (51.6%) of the electorate (Vanderleeuw, 1990).

New Orleans has a great variety of ethnic groups, two of the newest being Hispanics and Vietnamese. About 150,000 Hispanics live in the New Orleans area. The two largest groups are Hondurans, with perhaps 50,000 people, and Cubans, numbering about 12,000. No Hispanics have been elected to political office in the metropolitan area. In fact, Hispanics have yet to organize a grass-roots civic movement. During the 1990 elections in New Orleans, the first Hispanic to run for City Council was defeated soundly. More than 13,000 Vietnamese live in New Orleans. Most were brought here in 1975 when the Associated Catholic Charities assumed the responsibility of helping resettle Vietnamese refugees. Others migrated to be reunited with family or to search for the fishing life-style and warm climate of their homeland. The Vietnamese have been involved only marginally in the political life of the New Orleans community.

New Orleans's place in the hierarchy of American cities has changed considerably in 30 years. It is hard to believe today that New Orleans was the third-largest city in the country in the 1840 Census. Until the 1950 Census (when it was surpassed by Houston), New Orleans retained its status as the largest Southern city. New Orleans's Southern competitors are now such cities as Orlando (Fogelsong, 1989), Nashville, Memphis, and Birmingham (Perry, 1990a). The New Orleans metropolitan area has long since been surpassed in population and economic terms by such Southern metropolises as Dallas-Fort Worth (Elkin, 1987), Atlanta (Stone, 1989), Houston (Parker & Feagin, 1990) and, even Tampa (Kerstein, 1990). New Orleans is still a charming city with a beautiful architecture, outstanding cuisine, excellent music, and *joie de vivre*. In Noyelle and Stanback's terms, it is a "regional . . . diversified service center" (Noyelle & Stanback, 1983, p. 56).

INDUSTRIAL CHANGE

During the 1960-1990 period, the New Orleans economy had a tripartite base: the port, oil and related industries, and tourism. These activities represented more than 40% of the area's gross metropolitan product. Some observers would add government employment as a fourth part of the local economic base. Given the city's location near the mouth of the Mississippi River, the port was central to New Orleans's growth. In recent years, the port has faced problems of competition from other cities (e.g., Houston, Tampa) and has suffered decreases in the amount of tonnage passing through. The port provided 11% of the city's jobs in 1979. A steamship company executive, W. James Amoss (1985), notes that the number of workers who loaded and unloaded ships in the port dropped from 5,103 in 1970 to 2,088 in 1984. At the same time, the number of hours worked decreased from 7.51 million in 1970 to 2.96 million in 1984. Since most of this labor force is black, the effects upon the black community are especially severe, as blacks have lost relatively high-paying unionized jobs on the docks.

Since 1986 the Dock Board, the state agency that runs the port, has been headed by Ron Brinson, a dynamic young executive. Under his leadership the Dock Board has recovered from the brink of financial insolvency through strategic reorganization, has refined its marketing strategies and begun to pursue more aggressively a North-South trade pattern, and has begun a long-needed $187 million capital improvements program. The Board is committed to funding at least $87 million of this program itself, in conjunction with a $100 million grant from the state of Louisiana's Infrastructure Trust Fund. The Dock Board, unlike many other port authorities throughout the United States, has no source of operating subsidy and does not anticipate one in the near future. Therefore, the Board faces difficult problems in constructing and maintaining the kind of riverfront facilities it needs to match those of its competitors. Still, the Board continues to pursue its role as a catalyst for economic development in the New Orleans area. Realizing that the need for dockside labor will continue to diminish due to automation, the Dock Board is encouraging the creation of value-added activities to stimulate a broader economic role for the port.

The oil industry became an important part of the New Orleans economy after World War II, with the discovery and drilling of oil in the Gulf of Mexico off the Louisiana coast. As the oil industry developed, an industrial corridor emerged, extending from Baton Rouge (upriver from New Orleans) to the Gulf of Mexico (downriver from New Orleans) along the Mississippi River. Petrochemical industries are the most important component of this industrial corridor. The Chamber of Commerce estimates that one quarter of the jobs in the eight-parish metropolitan region relates to oil firms. Until recently, these corporations provided thousands of jobs in the central business district in their regional or exploration headquarters.

The drop in the price of petroleum and related products depressed the New Orleans local economy in the 1980s, while much of the rest of the United States enjoyed economic prosperity. The city had a double-digit unemployment rate for more than three years. Collapse of oil- and gas-related industries led to a population outmigration; well-educated young professionals were the most likely to leave the area.

As the local economy slowly recovered in the latter part of the 1980s, there was a change from petroleum-industry-related jobs to services jobs and tourism-related jobs. The major components of the area's service sector are as follows: retail trade (105,000 jobs), government (86,000), health care (34,600), wholesale trade (33,000), goods transportation services (27,400), finance and insurance (26,400), business services (25,500), administrative offices (20,000), professional services (16,000), private education services (15,000), hotels (14,000), and communications and utilities services (11,000) (Metrovision Team, 1990). The federal government has a large presence in the New Orleans metropolitan area.

Tourism plays a significant part in the rise of service jobs. New Orleans is a major tourist destination, and hospitality services generate more than twice their national share of jobs in the New Orleans metropolitan area. Area restaurants are especially important contributors (Metrovision Team, 1990). The move toward economic recovery is attributed in large part to an increase in tourism-related jobs. The number of convention delegates increased to more than 1 million per year after the opening of the new convention center in 1985. In addition, the number of hotel rooms increased from 10,000 in 1970 to 25,000 in 1985, in part as a result of the 1984 World's Fair. It is clear that tourism is now a dominant sector of the local economy, and this change has been encouraged by economic and political leaders. We must note, however, that many tourism-related jobs are in low-paying positions with high turnover rates. Because the tourist industry in New Orleans is cyclical, with fall and spring being peak times, it is necessary to expand staff during high-occupancy months and to reduce staff during other times. Many of the positions that need to be expanded and reduced require only minimal skills and education. Thus, although the number of jobs in the tourism industry is increasing, they do not offset the loss of jobs in the port and oil-and-gas, which tended to be higher-paying and with more secure tenure.

Manufacturing has played only a small role in the New Orleans economy. Interestingly, manufacturing has been part of the area's economic recovery in the shift from petroleum production and related construction activity. In 1986 the Department of Commerce reported about 47,400 manufacturing jobs in the metropolitan region, around 8.7% of total employment. Almost three quarters of the MSA's manufacturing jobs are in six sectors: ship building and repair, chemicals, food products, guided missiles and space vehicles, petroleum refining, and printing and publishing (Metrovision Team, 1990). The decision by a national discount chain to locate a major distribution facility in New Orleans will add more jobs in this sector and may stimulate additional activity (Bonnet, 1990).

THE RISE OF NEW COALITIONS AND MINORITY POLITICIANS

At the start of this era (1960), the mayor of New Orleans was DeLesseps ("Chep") Morrison. Morrison had been elected mayor in 1946, in a seeming victory for reform over the traditional Democratic machine forces that dominated New Orleans politics. During his early years in office, Morrison fulfilled some of the promises of reform. He created the New Orleans Recreation Department (NORD), which gained national recognition for its programs. The adoption of a new strong-mayor charter in 1954 was a major achievement of the Morrison years.

Morrison combined the uptown establishment and dissident regular Democrats into a new political machine, the Crescent City Democratic Association (CCDA). Morrison effectively used the all-black Orleans Parish Progressive Voters' League (OPPVL) to mobilize the black community on his behalf. Morrison's defeat of the incumbent machine mayor, Robert Maestri, was the first local election with significant black participation in many years. Blacks began to vote in substantial numbers after the abolition of the white primary in the early 1940s. The OPPVL continued to be important to Morrison throughout his years in office, especially at election time, but no blacks were allowed to become members of the CCDA. Thus, as Hirsch (forthcoming) argues, blacks were tied to Morrison personally, and the tie was a paternalistic one, not institutional.

In return for black support, Morrison dispensed favors and provided additional services to the black community. For example, he integrated the police force, and New Orleans constructed a substantial amount of public housing during his mayoralty. But, he did not share power, and he maintained many segregationist practices (Haas, 1974). For instance, he encouraged the creation of the Pontchartrain Park subdivision, an all-black area that encouraged the flight of the black middle class from the city's core and away from the newly developing lakefront area. It was Morrison's abdication of leadership in the 1960 school desegregation crisis in New Orleans that led to the rupture of the relationship between Morrison and the black community. The silence by Morrison and the city's business and social elite allowed archsegregationists to take the initiative, resulting in riots (Crain, 1968). The city's black leadership showed its displeasure with Morrison's behavior during the school desegregation crisis by trying to block his appointment as U.S. Ambassador to the Organization of American States (OAS). Morrison resigned as mayor in 1961, after his appointment to the ambassadorship.

Victor Schiro, a member of the City Council and a stalwart of the Democratic machine, replaced Morrison for the unexpired term. In 1962 Schiro won election in his own right, as he defeated State Senator Adrian Duplantier in the Democratic primary. Reelected in 1966, Schiro is known for the widening of Poydras Street, which led to the construction of the Superdome in the 1970s. Schiro was described in one analytical study (Kotter & Lawrence, 1974) as a reactive mayor who spent most of his time cutting ribbons and responding to such small, discrete issues as keeping an eye on trash in the streets. It is surely reasonable to say that he had the least black support of any mayor in this time span and that he was the last of the traditional white mayors in New Orleans.

The Voting Rights Act of 1965 had an immediate impact on the black political landscape in two ways. One was an increase in the black electorate, from 17.5% of the total in 1964 to 25.2% in 1966. The second impact was in the creation of black political organizations. The two major organizations formed at this time were SOUL (Southern Organization for Unified Leadership), a political offshoot of CORE), and COUP (Community Organization for Urban Politics), a more conservative group found in the 7th Ward downtown Creole stronghold. These groups were created to serve as intermediaries to broker the black vote and became convenient channels for white politicians to gain access to blacks (Hirsch, forthcoming).

In 1969 the major black organizations endorsed councilman-at-large Maurice "Moon" Landrieu for mayor. His opponent in the runoff, Jimmy Fitzmorris, refused to appeal to blacks as a group and ran as the white candidate. Landrieu won with 90% of the black vote and a minority of the white vote; black turnout was 75%. Landrieu's success in garnering the black vote ensured that the black political organizations would be viewed as major players whose influence would be felt in the highest levels of local government.

Landrieu was well known to blacks long before he ran for mayor. As a freshman state legislator in 1960-1961, he opposed (virtually alone) the segregationists at every turn during the school crisis. As a leader in the City Council (1965-1969), he responded sensitively to black needs. During the 1969 mayoral campaign, he endorsed Ernest "Dutch" Morial for councilman-at-large (an election Morial lost by only 5,000 votes).

Once in power, Landrieu worked hard to open opportunities for blacks through the political process. He increased the number of blacks in civil service jobs, from 19.4% in 1970 when Landrieu took office to 43% in 1978 when he left office (Hirsch, forthcoming). Landrieu named blacks to high appointed positions and provided significant patronage to blacks. During Landrieu's terms as mayor, a number of political offices were won by blacks, all of whom were Landrieu allies and as a direct result of his backing.

As Hirsch has suggested, "Landrieu was able to offer new opportunities to blacks without grievously antagonizing whites because he was cutting large new slices off an expanding economic pie" (Hirsch, forthcoming). This was possible because of Landrieu's skill at getting, for the first time in New Orleans, significant federal revenues. Much of these were antipoverty funds that he funneled through predominantly black channels, especially Total Community Action (TCA). State funds also were forthcoming, following

Edwin Edwards's election as governor in 1971 with the help of black voters. Especially important for the patronage that could be dispensed was the development of the Louisiana Superdome, opened in New Orleans in 1975.

Landrieu and Edwards supported the creation of Superdome Services Incorporated (SSI), a corporate umbrella under which many of their black allies could acquire economic resources and political leverage. Headed by SOUL leaders Sherman Copelin and Don Hubbard, SSI had 250 permanent employees and hundreds more for events (Hirsch, forthcoming).

In the 1977 electoral campaign, Ernest "Dutch" Morial, a state appeals court judge, was elected the city's first black mayor. Morial won by 5,471 votes, with 97% of the black vote (and a black turnout of 76%) and 19% of the white vote (mostly affluent uptowners), but without the support of most of the established black leadership. Most of the black political organizations saw Morial as a spoiler who did not appreciate political patronage; therefore, they threw their lot in with liberal white candidates (one of whom was Chep Morrison's son) who might be more forthcoming with patronage. Morial defeated the liberal whites during the primaries and then went on to the runoff and defeated Joseph DiRosa, a conservative white city councilman.

Morial felt he had run an independent campaign, and his early appointments reflected his emphasis on merit, not obligation. For example, some of his mid-level appointments were from Harvard's Kennedy School of Politics, where Morial participated in seminars before his inauguration. Most of the established black political organizations lost access to the kind of patronage they had enjoyed during the Landrieu years. Morial also tried to crack the entrenched bastions of white privilege and power by demanding free and open access to the public's business across-the-board.

Morial's political convictions and skills were tested early in his mayoral term, when he was forced to deal with the police department, one of the most abusive in the country. He brought in James Parsons, who had cleaned up the Birmingham force, as police chief; he broke the police strike of 1979; and he demanded an accounting of what happened during the Algiers shootings (when police officers on the pretense of looking for a police killer forced their way into an apartment in the Algiers section of the city and killed three people). As a result of the Algiers incident, the city created the Office of Municipal Investigation, Parsons was fired for not being aggressive enough in resolving the matter, and Warren Woodfork became the city's first black police chief. Morial succeeded in professionalizing the NOPD, opening it up and making it more accountable.

Morial asserted a primary role for the office of mayor (in the past the mayor was not considered one of the influentials in the city), and this brought him into conflict with the city's social and economic oligarchy. Since the turn of the century, members of the oligarchy have ruled from their positions on the many nonelected boards and commissions that govern such major areas as the port, the lakefront, sewerage and water, and the city debt. In 1979 Morial took on the Board of Liquidation, City Debt, which was self-perpetuating, by encouraging a lawsuit to end the Board's "racially restrictive membership." The judge ruled against the Board and nullified the last two appointments and ordered them to make a good-faith effort in considering a nonwhite (Hirsch, 1990). By the time Morial left office, most of the major boards and commissions had at least one black member. Throughout his term in office Morial bristled at the independence of the boards and commissions while he urged them to work in tandem with the city administration to improve the area's economy.

Morial tried to encourage the city to pay its own way and to break its dependence on federal and state aid. City employment dropped by one third during his administration, and he introduced a number of local revenue measures in an attempt to resolve the growing fiscal crisis. City employment dropped because of cuts in federal and state funds and because Morial's revenue-raising measures did not pass. Morial said that his greatest disappointment in office was his failure to come up with an adequate revenue base.

Morial's 1982 election campaign was a very bitter one that centered largely around what was termed his "arrogant personality." Morial won reelection with 99% of the black vote and the endorsement of most black political organizations, and 14% of the white vote (Perry & Stokes, 1987). In percentage terms, Morial received 53.2% of the vote to white candidate Ron Faucheux's 46.8% of the vote. The difference from his first election may be attributed to a decline in middle- and upper-income white support. After the election, Morial found himself besieged by the "Gang of 5" on the City Council that had the power to blunt his initiatives and to override his vetoes. In an attempt to regain greater control and to lose his lame-duck status, Morial twice tried unsuccessfully to change the two-term limitation on mayor in the city charter.

Perry and Stokes (1987) measure Morial's impact on blacks in the following ways: (a) Municipal employment—in 1977, blacks accounted for 40% of the municipal workforce; in 1985, they accounted for 53%; (b) Executive appointments—7 out of 12 (58.3%) of the department heads were black, the chief administrative officer was black; and Morial appointed the city's first black police chief; (c) Municipal contracts—two executive orders were passed that mandated the percentage of minority labor force participation in all city financed/administered projects and specified the goal of minority subcontracting participation. All told, blacks have made progress in obtaining a larger piece of the governmental and economic pie. The gains, however, are less impressive than those obtained in other cities.

BARTHELEMY'S TERM IN OFFICE

With Morial unable to run for a third term of office, the 1985-1986 mayoral campaign pitted two major black candidates against each other, City Councilman-at-large Sidney Barthelemy and State Senator William Jefferson (Feeney & Katz, 1985). The light-skinned Barthelemy grew up in the heart of the Creole 7th Ward. He became the first of his siblings to go to college and went on to get a master's degree in social work from Tulane University. He began his career in public service as head of the city Welfare Department during Moon Landrieu's administration. With the support of COUP, in 1974 he became the first black state senator since Reconstruction. He became the city's first black councilman-at-large in 1978 and was reelected in 1982. While on the Council, he waged a continual battle with Morial and later entered the mayoral campaign as the front-runner.

Jefferson grew up in northeast Louisiana in a poor, rural family. He graduated from Southern University and Harvard Law School. In 1976 he became the founding partner in what is still the city's most successful black law firm. In 1979 he was elected to the State Senate from uptown New Orleans, defeating a 16-year incumbent. While he and his law partners supported Morial in his 1977 mayoral campaign, in the 1982 campaign Jefferson ran against Morial and got trounced. When Ron Faucheux, a white candidate, entered the 1982 race, Jefferson was squeezed in the middle. The blacks stayed with Morial to make sure Faucheux did not win, and the whites went with Faucheux. Jefferson received only 7% of the vote, mostly from affluent whites from uptown. Shortly after the election, Jefferson and Morial reached a détente and shared a "peaceful coexistence."

In the 1986 campaign Morial backed Jefferson in what he viewed as his best chance to beat Barthelemy. Morial's support helped Jefferson run first in the primary among a field of three major candidates (the other candidate was Sam LeBlanc, a white attorney and former state representative, who ran as a White Hope candidate). In the runoff Jefferson lost to Barthelemy, 42% (67,698 votes) to 58% (93,049 votes). Jefferson won in the predominantly black wards, but Barthelemy took the white wards ("Unofficial Precinct Returns," 1986). Thus, Barthelemy won the election because he had the overwhelming majority of the white vote (85%) and a significant minority of the black vote (30%). Whites, who turned out to vote in slightly larger numbers than blacks, cast the pivotal votes in Barthelemy's election. In Barthelemy, white New Orleans had found itself a black mayor with whom it could live. The white vote for Barthelemy was seen also as a vote against Morial and his combative personality. Morial's endorsement of Jefferson helped Jefferson in the black community but probably cost him some white votes. With 51.2% of the electorate black, the results of the 1986 mayoral election suggest that the white vote played the same role that the black vote played in the 1960s and 1970s, when it was the minority component of the electorate. The issues in the campaign centered more around style than substance. Barthelemy's campaign theme was unity and harmony as he pledged "to bring this divided city together," to get people of all races and socioeconomic backgrounds to work together to improve New Orleans.

During his term in office the city has cut drastically the number of government workers and down-scaled city services, due largely to a decrease in funds from external sources. While the mayor is not blamed for setbacks from the state, national, and international economy, he has been criticized for lacking a coherent vision for where he wants to take the city in the 1990s. Without that vision the city has been left to drift, and other groups not controlled by City Hall, both public and private, have been left to chart the city's course. Though Barthelemy is still viewed as likable and cooperative, those traits alone are no longer seen as sufficient for governing the city. The administration has continued to operate in a crisis mode, assembling plans as issues arise, instead of anticipating problems and working potential solutions into an overall strategy. The Barthelemy administration has been plagued with management problems, in part because he assembled a team with little government experience. This inexperience has resulted in mishandled projects and strained relations with the City Council. Barthelemy's detached management style and his reliance on a small inner circle led to two major scandals early in his administration (Cunningham, 1990). The first involved allegations that his administration gave some businesses favored treatment in collecting city sales taxes. The second involved the dismissal of the director of the Housing Authority of New Orleans. In neither case was the administration culpable in wrong-doing, but it appeared inept in its management.

Barthelemy has been successful in distributing patronage. Among the dominant black political groups he has strengthened the fortunes of his organization, COUP, and maintained good relations with the SOUL group. He quickly moved to take control of such patronage-rich city agencies as the Housing Authority of New Orleans (HANO), the Regional Transit Authority (a law recently passed now enables the mayor to name five of the eight RTA commissioners), the New Orleans Exhibition Hall Authority (NOEHA), and the Aviation Board. A federal audit of the airport's contracts was critical of the noncompetitive process used by the Aviation Board in awarding professional service contracts. These contracts, worth millions of dollars annually, tend to go to friends and political supporters of Barthelemy and City Council members (Theim, 1989).

The 1990 mayoral election saw Sidney Barthelemy reelected with an electoral coalition very different from that of his 1986 campaign. Barthelemy was opposed by Donald Mintz, a white attorney involved in numerous civic organizations, including the chairmanship of the Dock Board. In 1989, the major question was whether former Mayor Morial would enter the race. Polls taken in October 1989, both for the candidates and an independent poll, showed Morial and Mintz in a dead heat, with Barthelemy trailing far behind. Ultimately, Morial decided not to enter the race. Morial's unexpected death from an asthma attack on Christmas Eve accelerated a move of individual and organizational support to Barthelemy (Makielski, 1990).

In the 1990 election, Barthelemy won reelection, capturing 55% of the vote to Mintz's 44%. The electorate in New Orleans is 54% black, and 69% of the city's electorate voted. Barthelemy's reelection followed the more usual pattern for black mayors: Emphasize black solidarity, get out the black vote, and obtain a reasonable minority of white support. Of course, in this election Barthelemy had a white run-off opponent. In winning, Barthelemy received 86% of the black vote and 23% of the white vote. Observers of the election agree that the most interesting departure of the election was the low black turnout. Specifically, Barthelemy received black support, but it was unenthusiastic, with voter turnout percentages of 40%-60% in black wards (Perry, 1990b; Makielski, 1990).

In discussing the rise of minority politicians, we must say a few words about the City Council. The New Orleans City Council is small, with only seven members. Five members are elected from districts, and two are elected at-large. The City Council first became majority black in 1985. The four to three black majority held in the 1990 city election, with all the black incumbents reelected. The changes in the City Council were not merely symbolic.

Another opportunity for a black politician was created by the 1990 retirement of U.S. Representative Lindy Boggs. Boggs's congressional district was the only black majority congressional district in the nation with a white representative. Boggs was replaced by the first black U.S. representative from New Orleans since Reconstruction, William Jefferson.

THE DEVELOPMENT OF NEW POLITICAL INSTITUTIONS AND PROCESSES

The City Charter has not been revised since the 1950s. Any effort to reform the charter might flounder on one of two controversial issues. First is the size of the City Council. Any effort to increase the size of the Council would be opposed by existing Council members, whose power would be diluted. This is especially the case in regard to planning and zoning issues, over which Council members have substantial power. Second is the length of the mayoral term. The present charter requires mayors to leave office after two terms although they may be reelected after a four-year interval. Any effort to change this would be caught up in reaction to the incumbent.

Revision of political institutions might include the parish government. Although the city of New Orleans and Orleans Parish are coterminus geographically, they do not have a complete governmental consolidation. The criminal justice area provides a vivid example. The city has a substantial police department, with more than 1,800 employees. The parish has a district attorney's office, a criminal sheriff's office with 400 employees and a civil sheriff's office, with 75 employees. Moreover, the School Board, with almost 9,000 employees, is a parish operation with substantial influence, patronage, and autonomy. Charges of political corruption and poor management have been rampant.

Efforts at governmental modernization have been prevented by the city's chronic lack of money. We have detailed Mayor Morial's fiscal struggles; Mayor Barthelemy has had similar problems, with layoffs of city workers and defeats of proposed taxes by the electorate. As Perry (1990b) points out, New Orleans electoral coalitions have broken down repeatedly on revenue-enhancing proposals, with consequent implications for the quality of life. In brief, New Orleans has suffered from cutbacks in federal and state funds in the last decade. This shortage is compounded by a high homestead exemption ($75,000) based in the state constitution, a system of assessors elected on a district basis, and a state legislature that thwarts city efforts at raising taxes. On tax questions, white homeowners consistently have opposed millage increases. In some cases, they have been joined by black voters and by segments of the white business community in opposition to taxes.

Faced with this situation, how has the city coped? Like other cities, New Orleans has shrunk city government (1989 saw 1,000 fewer employees than 1986). With no income tax and minimal property tax, there is heavy reliance upon sales tax (at 9%, one of the highest in the nation). We note that the city has made a major effort to develop its riverfront property in recent years. Anthony Mumphrey suggests that the city has turned to riverfront development as a substitute for annexation (Mumphrey, Wildgen, & Williams, 1990). With the city geographically constrained from annexation, riverfront development represents the only substantial opportunity to enhance the tax base. It is also a way of increasing tourism, a major economic generator for the city. In recent years, riverfront development has included Canal Place, an office/retail/hotel development; the Jackson Brewery and Riverwalk, two "festival marketplace" developments; and the New Orleans Convention Center and hotels. In September of 1990, an aquarium opened on the site of the old Bienville Street wharf, thus adding another tourist attraction to the city's riverfront development. The siting of the aquarium on an old wharf also represents the shifting uses of the downtown riverfront from port related to tourist oriented.

The way the city gets things done can be seen in a discussion of two nonprofit entities—the Friends of the Zoo and the Audubon Institute. These demonstrate the changing relationship between the public and private sector.

During Morial's administration an attempt was made to strip the mayor of his authority over the Audubon Park Commission (APC), which governed the uptown park, the zoo, and the white-only golf club (Hirsch, 1990). Under the Barthelemy administration, relations between the APC and the mayor's office are very smooth. Many of the APC's leaders were Barthelemy backers in the 1986 mayoral campaign. Some of APC's earlier goals are being realized, including privatization of some aspects of the APC and the exemption of zoo and park employees from the civil service. The impetus for both of these changes is the aforementioned riverfront aquarium, developed by the Audubon Park Commission and fully backed by Barthelemy.

Barthelemy has called the aquarium project the centerpiece of his economic development plan, a plan that stresses the primacy of tourism. The Aquarium Project is expected to do for the downtown riverfront what the Superdome did for the Poydras Street area in the 1970s, when it served as a catalyst for $3 billion of new construction. Stressing increased tourism, jobs, and economic vitality, the backers of the project—the Audubon Park Commission—pulled together a rare consensus of community support. That support included the entire City Council, the Business Council, the Chamber of Commerce, the NAACP and Urban League, the tourism industry, and riverfront developers. The small amount of opposition that surfaced had to do with the site of the project—along the riverfront on the edge of the French Quarter—rather than with the project itself. Preservationists were against the site because they felt its scale and the resulting traffic would harm the character of the Quarter; merchants within the heart of the Quarter (i.e., along Bourbon and Royal streets) feared the increasing competition from businesses along the riverfront.

New political institutions and processes have not extended to metropolitan institutions. Although the New Orleans metropolitan area is interdependent as an economic and social entity, the metropolitan fragmentation that is characteristic of other U.S. metropolitan areas is evident. The Regional Planning Commission (RPC), established in 1962, attempts to foster governmental cooperation and planning on areawide problems in Jefferson, Orleans,

St. Bernard, and St. Tammany parishes (i.e., New Orleans and the three suburban jurisdictions contiguous with the city). The RPC has only limited capacity and authority to deal with the problems. Relationships among governments in the metropolitan area often have been highly conflictual, with Orleans-Jefferson relations frequently embittered. In recent years, Orleans and Jefferson parishes have quarreled over the possible establishment of a city earnings tax on workers in New Orleans, the Regional Transit Authority, and street barriers between Orleans and Jefferson parishes.

Conflict and fragmentation are evident in economic activities in the metropolitan area. On an individual level, the consumer who purchases a new car in Jefferson Parish costs Orleans Parish money (and vice versa). On a collective level, businesses—both local and external—may choose one jurisdiction over another for their location. Cities and parishes within the region complete with each other for businesses. Political conflicts between city and suburban jurisdictions have occurred because of the changing demography in Jefferson Parish. In the future we can expect a further decline in the New Orleans population, stagnation in Jefferson Parish's (the closest suburban jurisdiction) population, and growth in such outlying parishes as St. Tammany (although this growth occurred in the earlier part of the 1980-1990 decade). The city still will have a clear black majority, and the suburban parishes will have a substantial black population despite the white flight.

New Orleans and the three major suburban jurisdictions (Jefferson, St. Bernard, and St. Tammany parishes) have worked together in a Regional Planning Commission (RPC) for two decades. The RPC, however, never has taken the assertive role on metropolitan issues that its equivalents elsewhere have. More recently, New Orleans and Jefferson Parish have had a series of disputes. One erupted when Mayor Morial attempted to impose an earnings tax on suburban commuters in the early 1980s. Other conflictual issues included transportation, and sewage and drainage. In 1987 Mayor Barthelemy confronted Jefferson Parish officials over the imposition of street barriers between the two parishes, and won. Overall, governmental relations between Orleans and Jefferson parishes are relatively calm. In part, this results from the low-key personality and style of the major current actors. One may also detect the beginnings of cooperative efforts to solve common problems, as evidenced in the effort to clean up polluted Lake Pontchartrain.

On February 18, 1989, David Duke, a former Ku Klux Klan leader and Nazi sympathizer, won a seat in the Louisiana House of Representatives from the New Orleans suburb of Metairie. Running as a Republican, Duke beat John Treen, a pillar of the Louisiana Republican establishment, by 227 votes. Duke skillfully used his legislative seat to push his platform: opposition to affirmative action and minority set-side programs, opposition to welfare, opposition to school busing, and no new taxes. In 1990 Duke parlayed the legislative seat into 44% of the vote in a statewide race against incumbent U.S. Senator J. Bennett Johnston. Despite the loss to Johnston, Duke is now clearly a statewide (and, possibly, a national) political force.

Duke's campaigns demonstrate a significant return to white racism in his district and in the state of Louisiana. At present, they have not meant much for the New Orleans metropolitan area. Perhaps suburban voters are more likely to express overt racist sentiments, especially when discussing crime in New Orleans. Duke's interests and orientations are statewide and nationwide. To this point, he has not directed his energy, attention, and charisma to city-suburban relations in the New Orleans metropolitan area.

CHANGES IN THE URBAN POLITICAL REGIME

Two major types of changes have occurred in this 30-year span. First are the changes in the New Orleans economic elite. In 1960 a small group of the old New Orleans social elite clearly controlled the local economy (Chai, 1971). It can be said that the economic elite is now more open to outsiders and to the inclusion of some blacks. Businesspeople who came into the community from outside the city now play a leading role in devising an economic strategy for the city's future. Blacks and women now serve on important regional boards.

The news media remain closely allied to the city's economic elite. The *Times-Picayune*, the city's only newspaper, is conservative and in basic agreement with the economic elite. A *Times-Picayune* endorsement is still a substantial influence in a local political contest. The major television station, WWL-TV (Channel 4), formerly owned by Jesuit-run Loyola University but now owned by the TV staff, is also supportive of the community's business elite. Labor is not a substantial force at the local level in New Orleans.

Curiously, the city has both welcomed and resisted political change. On the one hand, conservative white homeowners no longer dominate the city and its spending patterns. On the other hand, even a black leadership is reluctant to raise taxes for purposes of redistribution. Former Mayor Morial favored redistributive measures while the current mayor, Barthelemy, has emphasized patronage and favors for constituents. As Perry (1990b) points out, New Orleans has a problem putting together a revenue-enhancing coalition. Blacks seem to be no more supportive than whites of revenue enhancement measures. In recent years, voters of both races seem alienated and even churlish when faced with tax measures. Several tax referenda, including badly needed raises for police and fire employees, failed in the

1990 mayoral election, and a 1989 statewide tax reform measure failed to pass in New Orleans, as well as statewide (Miron, 1990). The result is that New Orleans still looks to the state and federal arenas for fiscal solutions.

REFERENCES

Amoss, W. (1985, May 25). The troubled port: What it needs to do to survive. *New Orleans Times-Picayune/States-Item*, p. A11

Bonnet, M. R. (1990). A case study of the decision by Pic-N-Save to locate in AMID: Location theory and economic development incentives in practice. Unpublished master's thesis, University of New Orleans.

Chai, C.Y.W. (1971, October). Who rules New Orleans? A study of community power structure. Louisiana Business Survey, pp. 2-11.

Crain, R. L. (1968). *The politics of school desegregation*. Garden City, NY: Doubleday.

Cunningham, L. (1990, January 21). Political veteran values teamwork. *New Orleans Times-Picayune*, p. A1.

Elkin, S. L. (1987). *City and regime in the American republic*. Chicago: University of Chicago Press.

Feeney, S., & Katz, A. (1985, February 3). The men who would be mayor. *New Orleans Times-Picayune*, Dixie Magazine, pp. 6-13.

Fogelsong, R. E. (1989, August-September). *Do politics matter in the formulation of local development policies: The Orlando-Disney World relationship*. Paper presented at the annual meeting of the American Political Science Association, Atlanta, GA.

Haas, E. F. (1974). *DeLesseps S. Morrison and the image of reform*. Baton Rouge: Louisiana State University Press.

Hirsch, A. (1990). Dutch Morial: Old Creole in the new South. (Working Paper No. 4). New Orleans: University of New Orleans, Division of Urban Research and Policy Studies.

Hirsch, A. (forthcoming). Race, rights and the re-emergence of black politics in 20th century New Orleans. In A. Hirsch & J. Logsdon (Eds.), *Creole New Orleans: Race and Americanization.*. Baton Rouge: Louisiana State University Press.

Kerstein, R. (1990, April). Tampa growth politics: Strains in the privatitic regime. Paper presented at the annual meeting of the Urban Affairs Association, Charlotte, NC.

Kotter, J. P., & Lawrence, P. (1974). *Mayors in action. Five approaches to urban governance*. New York: John Wiley.

Makielski, S. J., Jr. (1990). The New Orleans mayoral election. *Urban Politics and Policy Section Newsletter, 4*(1): 11-13.

Metrovision Team. (1990). Demographic & economic trends, New Orleans Region, April 26. New Orleans: KPMG/Peat Marwick.

Miron, L. (1990). *Corporate ideology and the anomaly of dedicated growth in New Orleans* (Working Paper No. 2). Division of Urban Research and Policy Studies, University of New Orleans.

Mumphrey, A. J., Jr., & Moomau, P. H. (1984). New Orleans: An island in the sunbelt. *Public Administration Quarterly, 8*, 91-111.

Mumphrey, A. J., Jr., Wildgen, J., & Williams, L. (1990). *Annexation and incorporation: Alternative strategies for municipal development* (Working Paper No. 1). Division of Urban Research and Policy Studies, University of New Orleans.

Noyelle, T. J., & Stanback, T. M., Jr. (1983). *The economic transformation of American cities*. Totowa, NJ: Rowman and Allanheld.

Parker, R. E., & Feagin, J. R. (1990). A "better business climate" in Houston. In D. Judd & M. Parkinson (Eds.), *Leadership and urban regeneration: The experience of twelve cities* (pp. 216-238). Newbury Park, CA: Sage.

Perry, H. L. (1990a). The evolution and impact of biracial coalitions and black mayors in Birmingham and New Orleans. In R. P. Browning, D. R. Marshall, & D. H. Tabb (Eds.), *Racial politics in American cities* (pp. 140-152). New York: Longman.

Perry, H. L. (1990b). The reelection of Sidney Barthelemy as mayor of New Orleans. *PS: Political Science and Politics, 18,* 2, 154-155.

Perry, H. L., & Stokes, A. (1987). Politics and power in the Sunbelt: Mayor Morial of New Orleans. In M. B. Preston, L. J. Henderson, Jr., & P. L. Puryerr (Eds.), *The New Black Politics: The Search for Political Power* (pp. 222-255) New York: Longman.

Stone, C. N. (1989). *Regime politics: Governing Atlanta 1946-1988.* Lawrence, KS: University Press of Kansas.

Theim, R. (1989, June 11). Mayor's backers make millions from airport. *New Orleans Times-Picayune,* p. A1.

Unofficial precinct returns. (1986, March 3). *New Orleans Times-Picayune,* p. A1.

Vanderleeuw, J. M. (1990, June). A city in transition: The impact of changing racial composition on voting behavior. *Social Science Quarterly, 71,* 326-338.

10

Denver: Boosterism Versus Growth

CARTER WHITSON
DENNIS JUDD

CONFLICTS OVER THE GROWTH ETHIC

A traveler approaching Denver on a clear day cannot help but be impressed by the cluster of shimmering skyscrapers jutting up from the plains, set in bold relief against a jagged cordillera. Except for the spectacular effect created by the mountain peaks, the scene would bring immediately to mind an observation by Carl Abbott about the new face of the Sunbelt cities: "Dallas, Los Angeles, Denver, Houston, Atlanta—the fast-growing American cities from one ocean to the other have built interchangeable cores. The uniform environment of high-rise offices, convention centers, sports arenas, and girdling freeways is an expression of shared values among urban leaders in our boom-town cities" (Abbott, 1987, p. 146). The idea of interchangeable skylines suggests that the political cultures of Sunbelt cities may be indistinguishable as well. Denver's style of politics, however, cannot be fit into such a uniform mold.

Like many frontier cities, Denver was founded with little more that plot stakes and a local newspaper (Dorsett, 1977). Boosterism in the form of unbridled local entrepreneurship and promotion has been central to Denver's politics ever since. Nevertheless, the city rarely has been dominated by the type of progrowth coalitions that have defined politics in other Sunbelt cities (Abbott, 1987; Feagin & Parker, 1990; Logan & Molotch, 1987; Stone & Sanders, 1987). Coalitions of this kind have proven to be temporary and fragile because of an abiding ambivalence about growth. During the past four decades in particular, Denver and Colorado boosters have aggressively pursued tourist and federal dollars. Concerns about air pollution, traffic congestion, and urban sprawl have been amplified because they manifestly endanger the "quality of life" so enthusiastically and publicly embraced by the residents of the city and the region. Thus, the local political culture is composed of both progrowth and proconservationist sentiments.

Unlike their counterparts elsewhere in the Sunbelt, Denver's civic leaders have not successfully built and nurtured stable progrowth coalitions that united public with private resources in the cause of urban revitalization. Even during those periods when an ideology of growth dominated the local political debate, the city government, with few exceptions, remained a minor player in efforts to bring economic development to the downtown. Most often, municipal leaders were carried along by the dynamics of the sustained prosperity of the region and failed to invest in political reform or public infrastructure to keep pace with population growth and its problems. Consequently, when the local economy turned sour in the early 1980s, they were ill equipped to respond. Their failure created the opening for a new generation of political leadership and a new style of politics.

In 1983 Federico Peña was elected Denver's 37th mayor with the support of a complex, broad-based electoral coalition. His campaign was built, in part, on promises to implement a comprehensive program to promote local economic growth. Downtown business interests provided crucial support. An appeal to ethnic residents and neighborhood groups, however, was the key to his electoral strategy. By promising to attend to the economic problems of downtown, as well as to the needs of neighborhoods, he succeeded in uniting concerns for growth with concerns for the local environment and quality of life. Bridging the gap has required a political dexterity that the mayor has only barely managed. Having survived two close elections, however, Peña has been able to implement a municipal agenda that is rare in the context of American city politics. In Denver, a coherent strategy of economic development exists in tandem with aggressive neighborhood-improvement programs.

BOOSTERISM AND GROWTH IN DENVER: 1948-1982

On a September afternoon in 1946, Denver mayor Ben Stapleton looked out at the city from his office window:

Traffic sped back and forth on Bannock St. and Broadway. Civic Center was alive with scores of persons converting the mall into a park. The trolley cars were jammed, because the automobile industry was still retooling from war production. These activities, visible from the mayor's office, were symptoms of a boom that would transform Denver from a quiet, conservative, fairly large city into a burgeoning metropolis with mushrooming suburbs and seemingly unsolvable problems. (Kelly, 1974)

Stapleton's solution to the problems he saw was disarmingly simple:

> "If those people would go back where they came from, we wouldn't have any problems here." (Kelly, 1974, p. 1)

The mayor's assessment probably was shared by other members of the business and civic elite who had ruled Denver for decades. By virtue of their control of the region's largest banks, three descendants of Denver's founders—Claude Boettcher, John Evans, and Gerald Hughes—shaped the city's development during the years between the short-lived expansion following World War I and the end of World War II. They preferred a slow but steady growth in the "clean industry," tourism (Dorsett, 1977, p. 220), and also lobbied for federal projects. They were not reluctant to engage in boosterism, but they did oppose new industrial development that might change the city. Their conservative credit policies favored the agricultural and livestock industries that earned the city its cow town reputation.

The conservatism of this tight-knit financial community matched the city's political torpor; in 1939 John Gunther, the famous chronicler of American political life, described the city government as "immobile . . . Olympian, impassive, inert. It is probably the most self-sufficient, isolated, self-contained and complacent city in the world" (quoted in Abbott, 1987, p. 127). Since becoming mayor in 1923, Stapleton's energies had been fully absorbed by a politics of cronyism and patronage, and over the years he had successfully resisted several attempts to reform the city government. He pleased his business friends by keeping taxes low and services at a minimum. Though he engaged in machine-style politics, Stapleton did not establish the rules and distribution networks characteristic of the party machines in the industrialized Northeast and Midwest. Rather than making municipal government central to the newcomers to the city, he kept it marginal.

The war effort brought far more industrial expansion to the city than the local governing regime wanted. The influx of population strained the city's housing supply and put pressure on its capital resources—its streets, bridges, sewers, water supply, and parks. The effects of fiscal and business conservatism were clearly visible in the shabby redbrick buildings of the downtown and nearby encroaching slums. To solve these problems, the newcomers, instead of leaving as Stapleton had hoped, fought for a political voice. In 1947 the widest vote margin in Denver's history elected 35-year-old Quigg Newton to the mayor's office.

In his campaign Newton charged Stapleton with cronyism and corruption, blamed the mayor for allowing the streets and parks to fall into disrepair, and accused him of catering to the financial elite that ran the 17th Street banking district. He promised to streamline city government, improve services, and promote new investment. Although he failed to get a reform charter past the voters,[1] he brought in a new generation of young professional administrators, got a sales tax passed, and implemented a competitive purchasing system and merit-based civil service (Abbott, 1987). His administration stressed "efficiency, economy, and automobiles" (Denver Office of Planning and Development, 1989). Newton and his successor, Will Nicholson, brought increased water supplies in anticipation of the population growth they predicted would occur (Dorsett, 1977).

Newton initiated a round of annexation that increased the boundaries of the city from the 63.4 square miles at the end of Stapleton's tenure to the 111.3 square miles that Peña inherited in 1983. This expansion enabled the city to expand its tax base while incurring only marginal increases in the cost of city services. In forging a progrowth coalition, Newton supported renegrade banker Elwood Brooks's loosening of credit at the Fifteenth Street Central Bank and also expressed public support for the downtown building plans of William Zeckendorf and the Murchison brothers. He enlisted the help of the Chamber of Commerce to promote new industrial and commercial development. By the end of the Newton/Nicholson years, even the *ancien regime* had been converted to the idea of growth, having discovered that "making money was not quite so unpleasant as had been feared" (Abbott, 1987, p. 129).

In the years following World War II, boosterism became a centerpiece of the region's politics, and it has yet to run its course. State, regional, and city leaders heavily promoted Colorado's tourism. In the 1960s the ski industry was one of the fastest-growing sectors in the state, and by 1975 more than 30 major ski resorts were open in Colorado. In 1980 tourism ranked fourth in the state's economy. So important is tourism that when ski resorts raise the price of lift tickets, it becomes front-page news all over the state (e.g., "Area Ski Resorts to Raise Prices," 1990).

Boosters also have been extremely effective at securing federal dollars. Even by the end of World War II, Denver had become the regional center for several federal agencies. In subsequent decades, delegation after delegation traveled to Washington to lobby for new facilities. When federal programs were consolidated into nine regional centers in 1967, Denver's designation as a center assured that federal employment would continue to be important to the local economy. By the end of 1975, more than 32,000 federal civilian employees worked in the Denver Federal Center or in other offices in the region (U.S. Civil Service Commission, 1977).

In the three decades after the war, Denver became firmly established as a federal city, a headquarters city for the oil industry, an administrative service

TABLE 10.1
Selected Demographic Indicators, 1950-1980

Indicator	1950	1960	1970	1980
Total City Population	415,789	493,887	514,678	492,365
Percent Denver SMSA	69.9%	53.1%	41.9%	30.2%
Percent Hispanic	5.4%	8.7%	16.8%	18.8%
Percent African-American	3.6%	6.1%	9.1%	12.0%
Percent Other Ethnic	0.8%	1.0%	1.8%	13.2%
Percent 18+ with college degree	9.3%	10.5%	12.4%	20.1%
Percent Individual Households	15.1%	25.1%	31.3%	43.7%

SOURCE: *Denver Data*, (4th ed., 1988, 10).

node for the Rocky Mountain region, and a favored location for subsidiary operations of national and international corporations. Like other growth cities of the West and Southwest, Denver's regional economy was not "structurally committed" to manufacturing (Noyelle & Stanback, 1984). Economic growth and diversification had an impact on employment patterns during the 1960s. Manufacturing employment grew by 24%, but the growth in services was many times faster—156% in professional services, 127% in business services, and 111% in education. In that decade alone, the number of jobs increased by 39.6% overall (comparatively, employment increased by only 19% in Cleveland during the same period) (Mollenkopf, 1983).

Though economic diversification may have insulated the region from the massive sociopolitical transformations of smokestack deindustrialization occurring elsewhere (Noyelle & Stanback 1984; Bluestone & Harrison, 1982), corporate activities and people were beginning to disperse out of the city and into the suburbs. Table 10.1 shows that from 1950 to 1980, Denver progressively became entangled in a suburban noose; in 1950, 70% of the metropolitan area's population lived in Denver, but by 1980 that share had declined to about 30%. Although the average educational level of the city's population improved with every census, the socioeconomic disparities between whites and minorities were considerable, and growing wider (Judd, 1986). By 1980 more than 30% of the city's population was Hispanic or black, compared with 9% in 1950. These emerging problems not only were neglected, but also they were exacerbated by a politics that focused exclusively on downtown development.

With remarkable similarity in design and purpose to renewal projects in other Sunbelt cities (Abbott, 1987), the Denver Urban Renewal Authority (DURA), a business-captured city agency, "led Denver to define urban redevelopment synonymously with downtown development" (Judd 1986, p. 181). Using urban renewal transfer funds to direct the federal bulldozer against the predominantly black neighborhoods bordering the aptly named Skyline Project in the central business district, the agency was content to leave the land as parking lots until developers snapped it up for high-rise construction. Clearance and demolition were so complete that by 1973 none of the 23 tallest buildings standing in Denver had celebrated a 20th anniversary; indeed, only a handful of downtown buildings had been built even in the initial boom of the middle 1950s.

Anchoring redevelopment near Cherry Creek on Denver's original settlement two blocks west of the downtown, 22 city blocks were razed for construction of the Auraria Higher Education Center, a complex housing the Denver Community College, Metropolitan State College, and the University of Colorado-Denver Center. That hundreds of Chicano families would be displaced and scores of architecturally significant Victorian houses would be leveled caused DURA and the city administration of Mayor Bill McNichols little concern (Abbott, 1987; Judd, 1986), although some ethnic and neighborhood leaders protested. The agency's independence from city government made its decision-making processes unaccountable to all except its business and developer clientele. Correspondingly, its independence also allowed City Hall to disavow responsibility for any of the projects it undertook.

The local economy was bolstered by the boom in energy prices that followed in the wake of the 1973 oil embargo. In a skyscraper boom that lasted from 1973 to 1982, 50 buildings were added to the downtown skyline. Between 1970 and 1978, employment increased by 34%, a rate second only to Houston's. In the 1970s, per capita income rose 27% and retail sales increased by an average 12.6% per year (Judd & Ready, 1986). The Currigan Exhibition Hall, which allowed the city to escape paying urban renewal matching funds for the Skyline Project (Abbott, 1987), and the $31 million Denver Performing Arts Center were constructed. Mile High Stadium and the McNichols Sports Arena became home to professional sports teams.

Between 1979 and 1982, more than $2 billion of private investment flowed into the Denver downtown north of Cherry Creek (Judd & Ready, 1986). By 1977 and without a millage increase, the 27 acres of the Skyline Project that had been DURA-created parking lots generating about $180,000 in taxes in 1967 were contributing $2.5 million a year to city coffers. Although less than 1% of the $1.27 billion total investment was spent on residential development (most of that for luxury condominiums), successful bond campaigns in 1967, 1972, and 1974 indicated continued majority support for redevelopment. These campaigns liberated DURA from a dependency on federal funds for its survival. The numerous bond issues passed during McNichols's reign earned him the name "Bill the Bondsman" even though he played only a marginal role in those connected with DURA and its activities.

Until the Community Development Block Grant (CDBG) program in 1974, redevelopment in neighborhoods meant the creation of vacant lots. The Community Development Agency (CDA) was established to administer projects undertaken when the CDBG grants became available. Since DURA already served as the clearinghouse for the downtown projects, the CDA became a neighborhood-oriented agency by default, all the more so since the mayor wanted to avoid getting embroiled in contentious fights over how to distribute CDA dollars. City Hall distanced itself from the neighborhood squabbles and the CDA even more completely than it had done from DURA and the developers. To insulate himself thoroughly from the CDA, McNichols created an independent Mayor's Advisory Commission (MAC) to decide how the monies would be divided among the competing claimants. The leaders of neighborhood groups often clashed with one another in meetings hosted by the MAC. The mayor studiously kept out of these battles even when his political allies asked him to intervene.

It is hard to prove that the McNichols administration followed a racially conscious policy in delivering city services, but if it had, the results would have been remarkably similar to the patterns that actually prevailed. In a study conducted in the early 1970s, Lovrich (1974) found significant differences in service priorities between Denver's Anglo, Chicano, and African-American voters. More than 70% of the Latinos and blacks in the study favored increased expenditures for police patrols, parks maintenance, recreation programs, and low-cost housing. These services did not arouse concern from even 40% of the white respondents. It is likely that the differing preferences among city residents were conditioned more by previous spatially distributed allocations than by any particular demographic characteristics. A member of Peña's administration observed that, "The streets used to be swept in the white neighborhoods and in north Denver because there were Italian crews. The housekeeping functions in the black neighborhoods and in the 'bad' neighborhoods were dreadful" (M. Sperling, personal communication, 1986). A representative of the Hispanic Chamber of Commerce said of the McNichols years, "It's like we weren't here" (S. Vigil, personal communication, July, 1990).

Denver holds a technically nonpartisan election for mayor every four years in odd-numbered years. Local and state party organizations' endorsements of particular candidates and the candidates' solicitation of those endorsements belie the nonpartisanship of the campaigns. If no candidate wins a majority in the May general election, a runoff is held in June between the two candidates with the highest vote totals.[2] An incumbent with less confidence in his record and his political clout than Bill McNichols would not have sought reelection in 1983.

McNichols's administration seemed to have fallen apart by the time declining energy prices brought recession to Denver in 1981. Having balanced the city budget for years by dipping into capital funds, the city's physical capital of parks, streets, sewers, and public facilities were in sorry shape, a state of affairs remarkably similar to the one that had faced the city after World War II, before the rebellion against the *ancien regime*. In 1981 DURA's political support collapsed when it announced plans to clear an area southwest of the downtown, populated by well-established small businesses. The agency was not able to offer any credible alternatives when a firestorm of criticism erupted. A few months later the tax collector was indicted for embezzling revenues. The newspapers revealed that improperly let contracts to McNichols's principals had cost Denver millions of dollars. A few months later the papers published lurid stories of police consorting with drug addicts and prostitutes. The final blow came from the skies in the same snowstorm that would spell doom for Jane Byrne's administration a thousand miles away in Chicago. After more than 2 feet of snow fell on the city over the Christmas holidays of 1982, city crews left some streets impassable well into the new year. Underneath these dramatic, public crises, a silent rot ate away at municipal government. It would have been exposed sooner or later. "If people knew that city receipts were literally being kept in cigar boxes," one Peña aide mused, "a little snow would have been the least of their worries" (M. Sperling, personal communication, 1986).

"IMAGINE A GREAT CITY"

The mounting problems in the city and the drumbeat of media coverage of small-scale corruption and incompetence encouraged several candidates to challenge the mayor in the spring of 1983. Voters were presented with the rare opportunity of selecting a mayor from a field of seven contenders. Though, unlike many other Sunbelt cities, Denver had never had slating committees, most mayoral contests had been limited to a choice between two white men with similar programs, outlooks, and backgrounds. This time the field was crowded. Among those entering the race against McNichols were the district attorney, a black state cabinet member, an Anglo state cabinet member, and former state representative Federico Peña, who had less than 5% name recognition going into the campaign. All of the candidates were professed Democrats.

McNichols's challengers stressed similar themes based on charges that McNichols was guilty of mismanagement, cronyism, and neglect of important problems. The Peña campaign organization put together pamphlets and campaign materials outlining detailed plans to coordinate economic and physical development and planning through regional interlocal cooperation. Peña promised to institute reforms to make city government more accountable to the people through community forums and a planning process that included neighborhood groups. To round out its campaign theme, "Imagine a Great City," the Peña organization issued position papers on the need to expand the airport, to preserve neighborhoods, to improve air quality, to create new jobs, and to bring a major league baseball team to town (Hero, 1987). Thus, Peña promised economic growth, tempered by concerns for the quality of life in the neighborhoods and citizen access to City Hall.

A Denver immigrant from a well-placed Galveston, Texas, family, Peña had been involved in the Chicano rights movement through his participation with the Mexican-American Legal Defense and Education Fund before his election to the Colorado House in 1978. But in his run for the mayor's office, Peña rarely commented on his Hispanic ancestry and did not stress his previous work on behalf of Latino causes. Similar to Tom Bradley in Los Angeles (Hahn, Klingman, & Pachon, 1976), Peña built a broad-based, issue-oriented liberal coalition by avoiding mention of potentially redistributive policies that might favor particular ethnic and racial groups and neighborhoods. The big campaign issues were designed to enlist support from environmentalists, feminists, and the aged, as well as minority voters. Although Denver's Chicano population had increased by 312% and its African-American population had increased 295% between 1950 and 1980, these groups together accounted for only 30% of the city's total population in 1983, a serious handicap for a candidate of Hispanic background. A door-to-door campaign that enlisted hundreds of volunteers was targeted especially at residents who had rarely voted in city elections. The nonvoters, who previously had supported Representative Pat Schroeder and Senator Gary Hart, were thought to lean moderately to the left. The Peña campaign also launched a massive registration drive in Latino and poor neighborhoods. If he could increase voter turnout to 70% from the usual 50%, Peña felt he would have a chance to win (T. Gougeon, personal communication, 1986; Hero, 1987).

On May 17, 1983, the highest voter turnout in Denver history gave Federico Peña 36% of the vote, District Attorney Dale Tooley 30%, "Mayor Bill" 23%, and the rest of the votes to the other candidates. In the runoff against Tooley, the Peña coalition was strengthened by organizational support from the black community, including the Black Roundtable and the

Colorado Black Chamber of Commerce. He also received endorsements from Schroeder, Hart, Governor Richard Lamm, the *Rocky Mountain News,* and the American Federation of State, County, and Municipal Employees, among others. There was a slight drop in support for Peña among Chicano voters between the general and runoff elections, a strong increase in support among African-American voters and Democrats, and more opposition among Anglos and voters with some college education (Hero, 1987). Allegations later surfaced that, in their zeal to support Peña, some improperly registered Chicano residents of Adams County crossed the city limits to cast their votes in the general and runoff elections (*Denver Post,* May 5, 1987:1B).

SYMBOLS AND SUBSTANCE

Mayor Peña refused to march in the 1984 St. Patrick's Day Parade because the sponsors prohibited lesbian and gay participation in the annual civic spectacle. Although the mayor was savvy enough to be out of town campaigning for presidential candidate Gary Hart, his absence made a point. Symbolism aside, the grassroots coalition of disaffected and dissatisfied voters, including lesbians and gays, expected substantive change.

The pace of change was frustrating to the mayor's supporters. Like Harold Washington in Chicago (Moberg, 1983), Peña proceeded cautiously and slowly in filling appointive offices. "When you're brand new as mayor and you have 60 appointments and you don't even have 60 people you can trust, it's very difficult to know who's going to turn out well and who isn't," (M. Sperling, personal communication, 1986) a longtime Peña supporter said in explaining the deliberations. Peña hired 10 African-Americans (16.4%) and 9 Latinos (14.8%). That these proportions were not greater may have resulted from a lack of solidarity among Denver's black and Hispanic organizations (Lovrich & Marenin, 1976). The presence of minorities in these numbers, however, was significant when compared to the McNichols administration, when less than 10% of appointed personnel were minorities. Expressing support for Peña but frustration at the pace of change, one unidentified source at the City Council described the Peña appointees as "probably the best educated, best informed people ever in city government and the least able to do anything" ("The Peña Administration," p. 1B).

Once in place, the administration moved quickly to consolidate and rationalize departments. "We're looking for long-term, systemic change as opposed to just changing a few faces" (A. Hayes, personal communication, 1986). The Denver Office of Planning and Development was created out of a merger of the Denver Planning Office (DPO), the Zoning Administration, CDA, and the Economic Development Agency—the successor to the now-defunct DURA. Clear responsibilities and paths of accountability were laid

out that ended back at the mayor's desk. The appointees' managerial inexperience, however, could not overcome resistance by career employees in the
vast city bureaucracy, and the appointees' political inexperience led to stormy
relations with the City Council. Councilwoman-at-large Cathy Reynolds
complained that the staff failed to "minimize their risks by talking to people.
There [was] a bunker mentality among Peña aides" ("The Peña Administration," p. 8B).

Although the administration sometimes gave the appearance of being
inept, projects that had lain dormant for years were now being completed. As
early as September 1984, *Odyssey West,* a local magazine oriented to the
black community, reported that substantial change was obvious in predominantly African-American neighborhoods. The neighborhood leaders interviewed by the magazine's reporter said they had noticed increased police
patrols in black neighborhoods, more City Hall assistance to black proprie-
torships, improved representation of African Americans in city government,
and new plans for neighborhood revitalization projects. Among most of those
interviewed, Peña was praised for remaining highly visible and accessible
through frequent "town meetings" held in various parts of the city and for
paying close attention to neighborhood concerns.

Although leaders in the Chicano community tended to echo these sentiments and cited similar changes, their support for Peña was less enthusiastic
than that expressed by black community leaders. Perhaps with expectations
of immediate rewards in exchange for an overwhelming display of electoral
support, many Latinos expressed dissatisfaction with the amount of access
to the mayor and a perceived inattentiveness to their needs. In fact, however,
Peña's administration was distributing money disproportionately to Latino
neighborhoods.

During the first term, the reorganized and consolidated Department of
Planning and Economic Development undertook 1,040 capital improvement
projects costing slightly more than $345.5 million in Denver's 77 statistical
neighborhoods. These projects included economic development loans to
small businesses, public housing rehabilitation, landmark designations, parks
and recreation maintenance, public facilities construction and rehabilitation,
public works projects, sewage and drainage construction and repair, street
maintenance, and neighborhood needs assessment studies. Table 10.2 shows
the distribution of these capital improvement projects.

Forty-seven neighborhoods received 72% of the total allocations. The
neighborhoods can be divided into three categories, as shown in the first
column of Table 10.2. The five neighborhoods with more than 60% Hispanic
population received a far greater proportion of the expenditures in relation
to their percentage of the city's population than any other areas, regardless

TABLE 10.2
Capital Improvement Allocations to Selected Denver Neighborhoods, 1983-1987

Neighborhood Characteristic	1983 Support for Peña	Number of Neighborhoods	Percent City Population	Percent Total Expenditure	Percent of Distribution by Category					
					Economic Development	Housing	Parks and Recreation	Public Safety	Public Works	Streets
> 80 Percent White	Against	8	12.17	6.89	3.40	19.88	0.50	28.41	3.71	9.70
	For	29	40.53	34.81	5.41	0.00	5.97	23.72	1.67	31.96
> 60 Percent Hispanic	Against	4	4.78	18.45	3.07	15.66	0.33	0.45	4.62	44.62
	For	1	0.44	9.61	2.26	NA	0.07	NA	78.03	1.51
> 60 Percent Blank	Against	0								
	For	5	6.84	2.18	9.28	19.17	0.21	0.02	1.37	0.69

SOURCE: Calculated from Capital Improvement Projects Maps, Denver Planning Office, 1987.

of the location of the neighborhoods or whether these neighborhoods supported Peña or not. The African-American neighborhoods fared the most poorly, although greater proportions of the total economic development and housing expenditures were allocated in these areas than in Latino neighborhoods. In white neighborhoods, electoral support for Peña appears to be the most straightforward explanation for allocation of capital improvement dollars.

Most of the visible accomplishments of Peña's first term were not dramatic; they could be overlooked easily. For example, fountains in city parks that had been turned off when McNichols depleted capital funds were ordered turned on. More streets and alleyways were rotomilled and repaved than in the previous years, and the trash was now picked up efficiently from city streets. Although these projects helped enhance the quality of life in neighborhoods, lasting improvements required "sound financial management and principles and [the creation of] adequate financial resources" (Denver Office of Management and Budget, 1984). Peña planned to use the city's available debt capacity to "pursue a higher threshold of economic viability," particularly in the downtown and nearby industrial strip in the Central Platte River Valley.

As soon as new financial management systems were in place to exploit the available debt capacity, however, the Peña administration was confronted by a decline in the city's creditworthiness. Remarkably similar to the experiences of Newark, Detroit, and Chicago after minority mayors were elected there (Davis, 1984), Denver's bond rating fell from AAA to AA. Although the city remained a good investment risk, the loss of its AAA rating reduced confidence in the mayor's ability to manage the city. Despite the high traffic in Larimer Square and on the downtown mall, Peña was not successful in luring a nationally prominent department store downtown. Peña infuriated developers when he opposed their plans to locate a larger convention center away from his preferred downtown site. While there was no disagreement that a larger hall was needed, the mayor entered the debate concerning its location, size, cost, funding, and design. Despite his promises to break ground for a new international airport, the mayor became entangled in lengthy negotiations with suburban counties for annexation rights and with airlines over their lack of cooperation in the project. When the airlines demanded veto power over the planning process, the mayor promised to go it alone.

Although the mayor was as anxious as the developers to promote downtown investment, he failed to coordinate a strategy that would combine public powers and private resources. Subsequently, development continued apace but with significant negative consequences. Between 1983 and 1987, 22.6 million square feet of office space and 15.1 million square feet of retail space were constructed throughout the city. By 1987 the office vacancy rate of 28.4% was second only to Houston's. Retail vacancies reached 16% (Frederick Ross Company, 1990). The problem of overbuilt capacity spilled well beyond the downtown into the whole city and the metropolitan area. Only 41% of the total office space was located downtown, but in the Denver Tech Center, a mixed-use complex of 2.5 million square feet located in the southeastern suburbs, space went begging. The downtown had plenty of competition, and it was clear that this would not abate even when the local economy improved.

The market in the housing sector was even softer. For years housing construction in the city had overemphasized shelter for the elderly. With Housing and Urban Development (HUD) subsidies, almost 4,000 luxury units were built in 19 projects between 1985 and 1987. Less than one quarter of these units could be expected to be occupied upon completion ("Retirement Home Builders," p. 1D). Vacancy rates also were high for other available apartments ("Metro Apartment Vacancies Fall," p. 45). When HUD dumped its inventory and offered homes at bargain rates, the local real estate market went into a depression.

It was in this atmosphere that, in early 1987, Peña announced his intention to seek reelection.

FROM OFFENSE TO DEFENSE

Peña's record in his first term offered some obvious targets to potential political rivals. City finances were in deficit. The state refused to authorize construction of the new convention center. The Federal Aviation Administration balked at the new airport plans. The police chief "who was responsible for doing the most to create the image of a no-nonsense, get-things-done mayor" ("Police Chief Coogan Quits," p. 1A) was forced to resign two weeks before the election because of his affair with a policewoman. A Peña backer, Lee Ambrose, said "Peña's knee jerk reaction [in firing the police chief] gave new life to old complaints that the mayor relied on the advice of advisors and doesn't think carefully before he acts" ("Peña Friends," p. 7A).

Five candidates opposed Peña in the election, one campaigning on the theme "Managing a Great City instead of Imagining One." Unlike the situation in his first election campaign, Peña emerged from the field a weak second. Don Bain, a local corporate attorney and political neophyte, received 41.3% of the vote to Peña's 36.8%. Three-term Democratic State Senator Dennis Gallagher captured 12.8% of the vote and was thus disqualified from the summer runoff. Hispanic and African-American voters failed to turn out for the primary in the same record numbers as in 1983.

The need for strong electoral support from Latinos and blacks forced a different kind of campaign on Peña than in 1983, a fact considered important by Bain's camp. In his attempt to rebuild the coalition that had elected him in 1983, Peña identified himself with minority causes and concerns more overtly than he had in the 1983 contest. Unlike the 1983 campaign, however, this one failed to involve grassroots volunteers and the Democratic Party organization, relying instead on Peña's appointees. With considerable help from the *Denver Post,* Bain attacked the Peña administration's record on city contracts let to minority-owned firms and on the level of minority hiring in city government, charging that under the Peña administration, minorities had been hired at a slower rate than ever before.[3] On the subject of minority hiring, Bain promised to recruit and hire "qualified" minorities and women who were "competent to manage, support, and communicate with the employees in their departments" (*Odyssey West,* March 1987:25). In reply to these charges, Peña pointed to his successful record. Under the circumstances—Bain's identification with downtown business and his race—these issues were not likely to catch fire.

Though both candidates stressed the need for downtown economic development, their neighborhood agendas were entirely different. In an apparent appeal to white middle-class and elderly voters, Bain said he would place priority on increasing police patrols. Peña countered by citing his record of supporting neighborhood revitalization. In this case his incumbency benefited the campaign, since he could point to tangible evidence of his administration's accomplishments, which he did at his frequent neighborhood appearances.

Although San Antonio Mayor Henry Cisneros came in early to campaign for Peña, local organizations, particularly several Chicano groups, were slow in giving Peña their endorsements. Public employee unions and 6 of the city's 13 councilors endorsed Bain. While polls showed steady gains for Peña during June, the Black Roundtable and the East Denver Ministerial Alliance, a Latino organization, waited until the week before the election to give Peña its support. Despite the closeness of the race, Gallagher, the third-place finisher, refused to endorse either candidate. From a 55% to 35% deficit in the polls on May 31, Peña's organization got out the vote in crucial neighborhoods on election day to win by 4,014 votes, a 51% to 49% margin.

During the runoff campaign, the mayor raised $887,491 and spent $1.22 million. His challenger raised $1.25 million and spent $1.35 million in the costliest election in the city's history ("Peña Raised $887,491," 1987). Turnout in 1987 was significantly lower than in 1983 *except* in those areas that had received lots of projects and money. A comparison of the 1983 and

1987 vote results showed that neighborhoods opposing Peña in 1983 were more opposed to him in 1987, while those neighborhoods that supported him in 1983 supported him by larger margins than in 1987.[4]

WEATHERING SIEGES

Denver's fiscal year begins October 1. Less than three months into his second term, Peña proposed a budget that would have eliminated the equivalent of 262 full-time employees from the city payroll. This was to be accomplished by laying off 130 employees and forcing five days of mandatory leave without pay. Two weeks earlier, Peña had shaken up his cabinet by giving key people new assignments. To keep the budget at the 1987 level of $387 million, he proposed drastic cuts in safety, public works, and recreation programs ("Peña Budget," 1987). The public reaction to these moves was swift and negative. Petitions were circulated for Peña's recall, and these were supported by large numbers of public works employees and other career civil servants. The recall effort fell short of the number of signatures required, but it left a legacy of bitterness within city departments.

Two days after his budget announcement, Peña named a new police chief, Aristedes Zavaras, a 43-year-old veteran with 21 years of experience on the force. Hispanics had lobbied hard for him. The other leading candidate was a 61-year-old African-American captain with 10 more years of seniority than Zavaras. Several black leaders charged that blacks had been betrayed ("Aristedes Zavaras," 1987). Neel Levy of the Colorado Black Chamber of Commerce (CBCC) vowed, "We won't be fooled again [by Peña]" (N. Levy, personal communication, July, 1990).

Peña also was making enemies among developers and businesspeople. One of the developers, billionaire train baron Philip Anschutz, joined with other members of the "Good Old Boys Network which had been disenfranchised from power when Mayor Bill was unceremoniously dumped" to form the Fifty-niners (Paige, 1989, p. 1B). This group included, among others, former Broncos owners, an ex-president of The Denver (a big retail store), a former newspaper publisher, an ex-Republican Party chairman, "and assorted other exes and formers." The group took out full-page ads in the *Post* and the *News* to announce their intention of keeping a watchful eye on City Hall and Mayor Peña.

Peña seemed unmoved by this threat. In the summer of 1990, when the landmark Central Bank, a 1911 Italian Renaissance Revival masterpiece of Denver architect Jules Jacques Benedict, was torn down over the mayor's protests, he withdrew city funds from the bank ("Historic Bank," 1990). In an open letter Peña wrote the bank president to say, "Our investment policies have been determined not only by the financial strength and integrity of the institution, but by that institution's willingness to contribute in a positive manner to the general well-being of the people of the city and county of Denver" ("Historic Bank," 1990).

In June of 1990 the convention center was completed, close to downtown exactly where the mayor wanted it. Located in a plaza across from the Currigan Exhibition Hall, the white steel and glass-fronted building held its first convention in July 1990 for 10,000 Christian booksellers who years before had promised not to return to Denver until it built a larger hall. Less than 1% of the affirmative actions jobs connected to the hall went to black-owned companies; the rest were awarded to Hispanic firms (Harding and Odbourne Law Associates, 1990). Although the CBCC cited this fact as evidence of discrimination by Peña's administration, other explanations are readily available. The Hispanic Chamber of Commerce was founded in 1979 and claims to represent 700 businesses, many of them engaged in construction, contracting, and trades. In contrast, according to a study commissioned by the CBCC, 72% of the 1,325 black-owned businesses in Denver were in selected services and retail, and only 11% of the enterprises paid wages to nonfamily employees (D. J. Miller and Associates, 1988).

The mayor failed to keep his promises to break ground for a new international airport before the second term when he became embroiled in lengthy negotiations with suburban counties for annexation rights and with airlines over their lack of cooperation in the project. In 1988, after making concessions to Adams County, Denver was allowed to annex 22 square miles for the airport. Construction began in 1989 on the first new international airport to be built in the United States since the mid-1970s, when the Dallas/Fort Worth airport was completed.

The mayor has been disappointed at his failure to lure a new top-of-the-market national department store to the downtown. When it opened in August 1990, the Cherry Creek Mall, 2 miles southeast of the CBD, had as its anchor stores Saks Fifth Avenue, Nieman-Marcus, May D & F, and Lord and Taylor's. The administration has decided that if it's located in Denver, it's still a good thing. Everything doesn't have to focus on downtown.

On August 14, 1990, the voters in the metropolitan Denver area passed a referendum to increase area sales taxes 1/10 of a cent to finance construction of a baseball-only stadium, in the event that the city is awarded a National League franchise in the next few years. By a 49.8% to 50.2% margin, the referendum failed within the boundaries of the city. Of the six counties in the SMSA, only Adams County, at the outer reaches of the metropolitan area, also failed to approve the measure. On the same day, Denver voters chose instead to fund a $100 million library bond issue. The lesson in this result may be hard to draw out, but it serves as a reminder that the politics of growth in Denver is unpredictable. The changeable attitudes of the general population are as difficult to predict as the growth agenda of Denver's elites are hard to classify.

DENVER'S UNCERTAIN GROWTH POLITICS

Growth politics in Denver cannot easily be characterized using accustomed categories from studies of other American cities. Progrowth coalitions have dominated the city's politics from time to time, but they have not been able to maintain a firm grip. They continued to rely on booster tactics to accomplish specific ends. The coalition forged in the late 1940s seems to have exhausted itself by the 1960s. Although downtown-centric renewal programs were implemented by the Denver Urban Renewal Authority, the city government mostly stood by the sidelines. Though business leaders had been free to pursue economic growth through their own efforts and later through the Denver Urban Renewal Authority, City Hall and the city's bureaucracies lacked the professionalism and the capacity to help very much. When economic problems came to Denver, it became obvious that the city's infrastructure was long-in-the-tooth and that its style of politics was worn out as well. Under Mayor Bill, city funds were kept in cigar boxes; the city did not even have computers.

Over the past several years Federico Peña has led the effort to build a local public administrative and political capacity for city government. His very visible accomplishments—the convention center, the airport—constitute a record that may be used to consolidate authority in the mayor's office. Whether Peña has constructed either a governing regime or a stable electoral coalition remains in question. Rivalries between downtown interests and the neighborhoods and among white, black, and Hispanic voters constantly threaten to tear it apart.

NOTES

1. Of the Western Sunbelt regional metropolises, Denver, along with Houston and Los Angeles, continues to operate under a strong mayor, unreformed form of city government.

2. In 1990 the U.S. Justice Department filed suit under the provisions of the Civil Rights Act of 1964 against the state of Georgia in the District Court at Atlanta. The plaintiffs alleged that this type of electoral process unfairly discriminates against minority candidates.

3. Bain claimed that the percentage of blacks and Hispanics holding top civil service appointments had declined between 1982 and 1986 from 1.7% to 1.4% for blacks, and from 1.2% to 1.1% for Hispanics. The mayor explained that, while the percentages had declined, the actual numbers of such appointments had indeed increased. He was not challenged regarding the implications that the size of city government had increased at an even faster rate than minority hiring (*Denver Post,* June 9, 1987).

4. While anyone's electoral support is ultimately a function of the votes received, it is relative to the votes the challenger also receives in a two-person contest. To establish a scale of Peña's support, the following formula was used:

Support = Votes for Peña/Votes for Challenger −1

This established an interval level measure that indicated the strength of the support for (Support > 0) or against (Support < 0) Peña in the neighborhoods for each runoff. Regressing 1987 results on 1983 results yielded a beta estimate of 1.24, with a two-tailed probability less than .0001. This indicated the hardening support or opposition in the various neighborhoods that overcame the decline in turnout for the 1987 election.

REFERENCES

Abbott, C. (1987). *The new urban America: Growth and politics in Sunbelt cities* (rev. ed.). Chapel Hill: University of North Carolina Press.

Adams County residents may have voted in '83 Denver election. (1990, July 18). *Denver Post,* p. 1B.

And then there were six: Denver's mayoral candidates seek office. (1987, April/May). *Odyssey West,* p. 24-27.

Area ski resorts to raise prices. (1990, July 18). *Denver Post,* p. 1A.

Aristedes Zavaras is new police chief. (1987, September 18). *Denver Post,* p. 10A.

Bluestone, B., & Harrison, B. (1982). *The deindustrialization of America: Plant closings, community abandonment, and the dismantling of basic industry.* New York: Basic Books.

Davis, T. (1984, July). Black mayors: Can they make the difference? *Mother Jones,* p. 32.

Denver Office of Management & Budget. (1984). *Preface to the 1985 Budget* (p. 1). Denver: Author.

Denver Office of Planning and Development. (1989). *1989 Comprehensive Plan.* Denver: Author.

Dorsett, L. W. (1977). *The Queen City: A history of Denver.* Boulder, CO: Pruett.

Feagin, J. R., & Parker, R. (1990). *Building American cities: The urban real estate game.* (2nd ed.). Englewood Cliffs, NJ: Prentice-Hall.

Frederick Ross Company. (1990). Retail and office square footage construction and occupancy report. Denver: Unpublished data.

Hahn, H., Klingman, D., & Pachon, H. (1976, December). Cleavages, coalitions and the black candidate: The Los Angeles mayoralty elections of 1969 and 1973. *Western Political Science Quarterly, 55,* 507-520.

Harding and Odbourne Law Associates. (1990, June). *The Disparity Study.* Denver, CO: Author.

Hero, R. E. (1987, March). The election of Hispanics in city government: An examination of the election of Federico Peña as mayor of Denver. *Western Political Science Quarterly, 40,* 95-103.

Historic bank loses fight for survival. (1990, July 11). *Denver Post,* p. 1A.

Judd, D. (1986). From cowtown to Sunbelt city: Boosterism and economic growth in Denver. In S. S. Fainstein, N. Fainstein, R. C. Hill, D. Judd, & M. P. Smith, *Restructuring the city: The political economy of urban redevelopment* (rev. ed.) (pp. 167-201). New York: Longmann.

Judd, D., & Ready, R. L. (1986). Entrepreneurial cities and the new politics of economic development. In G. E. Peterson & C. W. Lewis. (Eds.), *Reagan and the cities* (pp. 209-247). Washington, DC: The Urban Institute Press.

Kelly, G. V. (1974). *The old gray mayors of Denver.* Boulder, CO: Pruett.

Logan, J. R., & Molotch, H. L. (1987). *Urban fortunes: The political economy of place.* Berkeley, CA: University of California Press.

Lovrich, N. P. (1974, December). Differing priorities in an urban electorate: Service preferences among Anglo, Black and Mexican American voters. *Social Science Quarterly, 55,* 704-717.

Lovrich, N. P., & Marenin, O. (1976, June). A comparison of black and Mexican American voters in Denver: Assertive versus acquiescent political orientations and voting behavior in an urban electorate. *Western Political Science Quarterly, 29* 284-294.

Metro apartment vacancies fall below 10%. (1990, July 19). *Rocky Mountain News,* p. 45.

Miller, D. J., and Associates. (1988, March). *A study to develop a plan to address the economic growth and development of the Denver black community.* Denver: Author.

Moberg, D. (1983, October). The man who wants to break the mold. *Chicago,* 170-182.

Mollenkopf, J. H. (1983). *The contested city.* Princeton, NJ: Princeton University Press.

Noyelle, T. J., & Stanback, T. M., Jr. (1984). *The economic transformation of American cities.* Totowa, NJ: Rowman and Allanhead.

Paige, W. (1989, November 26). The good old boy network. *Denver Post,* p. 1B.

The Peña administration. (1987, March 22). *Denver Post,* p. 1B, 8B.

Peña budget to eliminate 262 jobs. (1987, September 16). *Denver Post,* p. 1A.

Peña friends believe Coogan resignation not handled well. (1987, May 7). *Denver Post,* p. 7A.

Peña raised $887,491 in campaign. (1987, September 2). *Denver Post,* p. 1B.

Police Chief Coogan quits over affair with policewoman. (1987, May 1). *Denver Post,* p. 1A.

Retirement home builders saturate Denver market. (1987, September 14). *Denver Post,* p. 1D.

Stone, C., & Sanders, H. T. (1987). *The politics of urban development.* Lawrence, KS: University Press of Kansas.

U.S. Civil Service Commission. (1977, March). Employment Trends as of January 1977. *Monthly Trends.*

11

Houston: Administration by Economic Elites

ROBERT E. PARKER
JOE R. FEAGIN

HOUSTON IS A BIG CITY where the most important political and economic decisions are made by a small number of powerful actors. As Shelton, Rodriguez, Feagin, Bullard, and Thomas (1989) have observed,

> In the mid-1980s the Houston Chamber of Commerce functioned as a type of local government. (In fact, the president of the Houston Chamber of Commerce earns a higher salary than the mayor.) It commissioned many of the planning studies of Houston's social and economic problems, including an important study on how to improve the economic diversity of the city. And the chamber has been active in setting up organizations to boost the city's economic image and to recruit new types of industry to diversify the city. (1989, pp. 134-135)

In short, the citizenry of Houston is more affected by the local business elites (now nestled in the Greater Houston Partnership) than by elected officials. As that suggests, much of Houston's "planning" is done by private agents. To speak of political change in Houston between 1960 and 1990 is to speak mainly about changes in the elite structure of the city. The potential for more traditional forms of political contest exists in Houston, particularly given the growing minority presence there, but it has not yet been realized.

BOOM IN THE SUNBELT

During the 1960 to 1980 period, the employment growth rate in the Houston area averaged 5.4% a year, compared with a rate of 2.8% for the United States. Between 1974 and 1981, nearly 700,000 jobs were added to the area's economy as nonfarm payroll employment increased 75%. Per capita income among Houstonians also was advancing faster than the national average during this time, a gain of 160% for Houston workers, compared with 108% for all metropolitan employees (Greater Houston Chamber of Commerce, 1989a).

It would be difficult to overstate the centrality of oil and gas to Houston's economy. The city has long been at the center of an international oil industry. In the 1980s, roughly 35% of the jobs in Houston were tied directly to the oil-gas industry, with perhaps another 20% connected indirectly. In the past decade fully 70% of Houston's large plants (those with 500 or more employees) were linked to the oil and gas industry. More than half of downtown office space was occupied by energy-related corporations, with much of the rest utilized by the banks, law firms, and other companies that service oil-gas companies. Also consider that 11 of the 20 largest Texas firms were oil and gas companies, 8 of which were headquartered in Houston, and that 34 of the 35 largest oil companies had major office and plant facilities in the greater Houston area. Finally, an estimated 400 major oil and gas companies were operating in the metropolitan area in the 1980s, together with hundreds of such support firms as drilling contractors, supply companies, and other oil-related businesses. In short, the oil industry and its spin-offs dominate the Houston economy.

Another central prop to the Houston economy is the Port of Houston. The value of foreign trade through the port increased from $2.4 billion in 1970 to $23 billion in 1980. In 1988 more than 64.5 million tons of import and export cargo moved through the Port of Houston, making it the leading U.S. port in foreign trade (Port of Houston Authority, 1989). The Johnson Space Center (JSC) of the National Aeronautics and Space Administration (NASA) is the major space-defense complex in Houston. By 1990 the space center employed 3,250 federal employees and 9,650 private-sector contractors.

Since 1962 the city economy has been stimulated directly by more than $40 billion in taxpayer-provided monies from the space center. According to John Brock, president of economic development for the Greater Houston Partnership (see fuller discussion of the GHP below), the impact might be 10-fold—maybe even 20-fold—on the general economy in Houston. The JSC spends nearly $4 million each working day in Houston and obviously creates substantial ripple effects. In addition to the strictly financial impact of the center, the JSC has been useful as a public relations tool for the city and has served as a focal point for other aerospace companies, such as Grumman Corporation and McDonnell-Douglas, to locate. Finally, the JSC is the single major source of research-and-development funding.

Although Houston recently has begun to recruit high-technology companies intensively, as of the early 1980s only three such companies were having a significant impact on Houston's economy: Compaq, Texas Instruments, and Zaisan, Inc. Of greater consequence to the local economy is Houston's high-technology medical industry. The components of the 235-acre Texas Medical Center in the early 1980s included 29 hospitals and research institutions. In all, the Center is estimated to produce more than $1.5 billion in economic effects. Finally, Houston's economy gets a relatively minor stimulus from federal spending on research and development. In recent years the Houston area has received just a fraction of the R&D dollars received by private companies in such cities as Los Angeles, Washington, San Francisco, and Boston.

HOUSTON'S DECLINE

Houston has an enduring history of economic growth, with its reputation as the "city that never knew the Great Depression." But the economic downturn during 1982-1986 demonstrated that Houston was not economically invincible. Nonfarm payroll employment peaked at nearly 1.6 million in March 1982 and then posted declines in 16 of the next 17 months. The city was hit hard again when the unemployment rate rose to 10.3% during 1986. By the beginning of 1987, the city of Houston had 221,900 fewer jobs than in March 1982.

In large measure, Houston's decline reflected trends in the world oil economy; OPEC production and the discovery of major new sources of crude, such as those in Great Britain's North Sea, have dampened oil prices as the city tries to reemerge from its recession. The city's over-reliance on the oil and gas industries generated major economic troubles when international oil prices tumbled in the mid-1980s. Another international factor, currency fluctuations, particularly those affecting the U.S. dollar and the Mexican peso, also hit the Houston economy hard in the early and mid-1980s. In sum, Houston is an international oil city, as much affected by decisions made by corporate and political leaders operating outside the United States as by decisions made within the United States.

As 1990 approached, several key sectors of the Houston economy remained in the doldrums. Energy was one such industry. A survey by the *Houston Business Journal* (Graczyk, 1988) found that 20 of Houston's 50 largest public companies lost money in 1987. Of these 20, 12 were energy related. Because of the weak price of oil, energy companies continued to lay off workers while the rest of the country was in the midst of a long economic expansion. For example, following the restructuring of their operations, Arco reduced its number of domestic employees by 23%, Phillips Petroleum cut its workforce by 7%, and in the fall of 1989 Occidental Petroleum announced it would decrease its domestic work force by 900 workers, or 20%. Staff reductions in Houston were integral to the reorganization plan. The energy industry, after having seen oil prices per barrel rise to more than $30, saw them drop to $10 in 1986. In the spring of 1990, the price for West Texas Intermediate oil (the U.S. benchmark crude) fluctuated between $16 and $21 per barrel. The oil patch has improved since the mid-1980s, but many companies already have gone bankrupt waiting for prices to stabilize, and still others will not explore for oil unless the price is consistently above $20 per barrel. Recently, of course, the price for a barrel of oil has jumped. In early October 1990, the price for a barrel was fluctuating around the $39 level in reaction to the Persian Gulf conflict, but the jump in oil prices does not appear to be having a profound impact on Houston's oil-based economy. Some believe it will be a temporary jolt, while others point to oil field technology that enables operators to extract additional oil from existing wells, rather than creating demand for new wells and oil field activity.

Another barometer of the oil industry's health is the number of working oil and natural gas rigs. In 1989, just 100 more active rotary rigs were operating than the postwar record low of 686 set in 1986. By contrast, more than 4,000 rigs were operating in late 1981. This situation has been altered somewhat by the situation in the Persian Gulf, such that by mid-September 1990 about 1,040 active rigs were operating (N. Rodriguez, 1990). In late 1990, the price for a barrel of oil was approaching $40 as the Gulf War loomed. Shortly after the onset of the war, oil prices declined by more than $20, the most severe single-day fall in history.

The real estate and construction industries also have resisted recovery. There is a particularly serious obstacle to Houston's full economic recovery, given the centrality of real estate and developers in the Houston economy. For example, Houston is home to some of the largest diversified U.S. developers, including Gerald D. Hines Interests and Century Development Corporation. With increased deregulation of the financial industry, at least 15 banks were opened by real estate developers, most of whom went into business to facilitate development activities. At least 3 of these went bankrupt during the oil-related decline of the 1980s.

The mid-1980s oil-gas recession produced high vacancy rates for industrial plants as companies reduced leased office and commercial space. Between 1978 and 1982 the vacancy rate in Houston was lower than the national average, but by 1983 it was rising dramatically. In 1984 the vacancy rate in Houston office buildings was running four times the rate normally considered profitable. By 1986, the vacancy rate had climbed to one quarter of all office space, and by mid-1987 the figure was approaching one third. Although occupancy rates have increased of late, the office vacancy rate remained stubbornly high at 27% in early 1990.

The real estate industry's low occupancy levels are particularly pronounced in suburban areas and in big office projects. New nonresidential contracts for all of 1987 reached $1.54 billion, down 21% from the previous year. Even that figure would have been lower were it not for public-sector spending. The city has 55,000 vacant lots, and while the price of homes has shown healthy advances, the $71,000 median price (in mid-1990) reflects earlier widespread losses in the value of property. Finally, the city continues to endure the burden of numerous insolvent financial institutions. Of the 10 largest failures involving Houston financial institutions, 6 occurred between March 1986 and May 1987. As with the JSC, it will be federal dollars that intervene in this free-enterprise city to achieve economic stability. Much hyperbole has surrounded the current modest recovery in Houston, but scant attention has been paid to the public role in nurturing it.

Other economic indicators signaled a recovery as the 1980s ended. In addition to the creation of 44,000 new jobs in 1988, Houston in early 1989 had a manufacturing employment growth rate exceeding 4%, an unemployment rate of 6.6%, an inflation rate of 3.7%, and a foreclosure level 37% below that of the previous year (Houston Economic Development council, 1989). Reflecting on the modest economic barometers pointing to recovery, Brian Levinson, spokesperson for the Greater Houston Partnership, employed stereotypically pre-bust Texan vernacular: "We not only survived it, but we did so more quickly and more powerfully than most people expected. In many parts of the country this kind of turnaround would take years" (Belkin, 1989, p. 25).

POPULATION CHANGE

For decades, Houston and the state of Texas have been the destination for many seeking jobs and a better standard of living. Between 1850 and 1980, the city grew from 2,400 people to almost 1.6 million, while the metropolitan area grew from over 18,000 to nearly 3 million. Between 1960 and 1980, Houston experienced a high pace of net in-migration, gaining about 30% in population in each of those decades. Houston's incorporated area grew from 938,000 in 1960 to 1,594,000 in 1980. By 1983 Houston had surpassed Philadelphia to become the fourth-largest city in the nation. The census takers now report the city's population to be 1.6 million, while the metropolitan area is more than double that much.

The rapid-growth trend slowed significantly as the region's economic fortunes turned. In 1982 the net in-migration was 417,000 for the whole state, but by 1986 the figure had dropped to fewer than 5,000 people. This demographic switch was significantly more acute in Houston than statewide. The loss of jobs reduced net migration to Harris County to less than 7,000 in 1984 from nearly 49,000 in 1981. And 1985 began the first of four consecutive years in which out-migration would exceed in-migration. During that period, Harris County lost an estimated 125,000 people (GHCC, 1989b).

Black migration to Houston began in earnest during and after World War II. Between 1940 and the early 1980s this migration markedly increased the size and number of Houston's black residential areas. During this period Houston's black population increased from 86,300 to 444,250. Older black communities in the 3rd, 4th, and 5th Wards grew dramatically, sometimes spilling over into adjacent white communities.

Over time, the geographic center of the black community has shifted. In the 1950s and 1960s the center of the black population moved northeast of downtown to the 5th Ward. The Lyons Street area was called by some one of the major black cultural centers in the United States, with many jazz clubs and businesses that catered to a black clientele. Later, much of the vibrancy that characterized the 5th Ward was destroyed by highway developments. Assisted by new waves of migrants, the black population shifted back to the 3rd and 4th Wards.

Some residential desegregation did occur in formerly all-white residential areas in the 1970s, and some black middle- and upper-income Houstonians moved out of the traditional areas during the next decade. A study in the mid-1970s found the 3rd Ward to be a high-density area with 51,300 residents, 91% black and 3% Mexican-American. In the mid-1980s three quarters of black Houstonians resided in predominantly black areas, which are generally segregated and distinctive. Recent estimates place the number of African Americans at 750,000-800,000, or about one fifth of the metropolitan population.

In addition to African Americans, Houston has other significant ingredients in its racial and ethnic mix. Particularly noteworthy has been the growth of the Hispanic population. According to U.S. Census Bureau data, by 1985 the number of Latinos in the Houston metropolitan area was just 7% below that of the African-American population (N. Rodriguez, 1990). It is now widely believed that the Hispanic population has matched or may even be larger than the black population.

In the 1960s and 1970s most Hispanic immigrants were undocumented Mexican workers, mostly temporary migrants who worked in the United States. In the 1980s the aspirations of people to escape conditions in certain Central American countries produced an exodus of workers to Houston.

The pace of recent Latino immigration has been relatively rapid, transforming the social landscape in some parts of the city. For example, one elementary school in the Galleria area went from 54% Anglo in 1980 to 92% Hispanic in 1986 (N. Rodriguez, 1990). In the mid-1980s a local immigration agency estimated that Houston had 100,000 Salvadorans, 30,000 Guatemalans, and 10,000 Hondurans, Nicaraguans, and Costa Ricans. In recent years Houston has become one of the largest Salvadoran communities in the world.

Official statistics in the mid-1980s placed the number of Hispanics in Houston at 250,000, with another 150,000 outside the city limits. Most observers believe these are conservative estimates. A marketing survey in the late 1980s put the number of Hispanics in the "Houston market" at 706,000, a figure that ranks the city as the seventh-largest Hispanic market in the United States. Several lesser-known immigrant groups have added layers of texture to Houston's racial and ethnic composition. For example, several hundred Garifunas, black-Caribs from Honduras, have settled in the 3rd and 5th Wards. The Garifuna stem from Latin ancestry but are racially black. Recent immigration to Houston also has brought approximately 2,000 Mayans from two provinces in Guatemala. N. Rodriguez (1990), noting the arrival of these new Latino groups, says the Hispanic population of Houston is 750,000 to 800,000.

HOUSTON'S BUSINESS LEADERS

Since its inception, Houston has been dominated by powerful business elites. Early on, cotton merchants were central, and then came leaders from oil, powerful law firms, banking, and real estate. Many cities have had significant working-class local representation at some point, but Houston has almost no history of such activity. In the late 1930s a highly cohesive group of business elites emerged, a group of powerful corporate executives known as the "Suite 8F crowd." They were a particularly significant force through the 1960s. By the mid-1970s their impact was largely diluted, but traces of this group's influence can still be found in Houston today.

One of the most influential business leaders was Jesse Jones. Not only was he Houston's first major property developer, Jones also had major capitalist interests in banking, oil, and communications. He was one member of the "Suite 8F crowd" who served in the government, as chair of the Reconstruction Finance Corporation and later as secretary of commerce. He was also a key actor in persuading Gulf Oil to relocate to Houston. Other important members of the "Suite 8F crowd" included James A. Elkins, Sr., Herman and George Brown, Gus Wortham, James Abercrombie, and an assortment of allied business people. Elkins helped found what is today one of the nation's largest law firms (Vinson and Elkins). At the local level, Elkins reportedly was central in shaping mayoral politics until the 1960s. Herman and George Brown have had an impact on city politics from the 1920s to the 1980s. Herman Brown formed what would become Brown and Root, a major construction firm heavily dependent upon large-scale projects subsidized by the federal government. Gus Wortham founded American General Insurance Co., one of the 20 largest insurance companies in the United States, and was instrumental in integrating the "Suite 8F crowd" into larger business and community interests and organizations. Another member, James Abercrombie, established Cameron Iron Works, which became one of the world's leading oil tools manufacturing firms.

The leadership position of the "Suite 8F crowd" was weakened significantly by the deaths of core members and the influx of more than 150 company subsidiaries, divisions, and headquarters between the 1960s and the late 1970s. In reference to the demise of the "Suite 8F crowd," Suro observes that "the city's business leadership is far too large and heterogeneous to fit in a single hotel suite, and diversification has replaced oil as the corporate byword" (Suro, 1989, p. 6). Others have argued that since the 1970s, the "Suite 8F crowd" has been replaced by "a more expanded oligarchy in which the key institution is the Chamber of Commerce . . ." (Davidson, 1982, p. 278). Since the 1970s, the executives named to the Board of the Houston Chamber of Commerce have come from a variety of major corporations, including industrial firms, banks, the media, law firms, and insurance and construction firms.

The emergence of corporate executives on the board indicates the important role played by multinational corporations in Houston in the 1970s and 1980s. Kenneth Schnitzer, Century Development Corporation's top executive, is representative of the increasing significance of major developers in Houston's business leadership. Other developers, such as Gerald Hines, also became members of the Chamber's board of directors. While the importance of the "Suite 8F crowd" has diminished, several of their representatives still can be found among Houston's business elites. Despite the diffused structure, in the 1980s just a few Chamber members could determine the fate of major development projects.

A PERIOD OF TRANSITION

The 1970s was a period of transition for Houston's business elites, a time of significant expansion in the representation of the Chamber's governing board of top executives. Despite the changing membership, the primary objectives have stayed the same: the protection, enhancement, and expansion of business investments. From 1960 to the present the "Suite 8F crowd" and later the Greater Houston Chamber of Commerce have done much of Houston's goal setting and planning and provided many of the city's elective officials.

By the 1980s the transition to a more expanded administration and corporate influence was nearly complete; mayors, city councils, and planning commissions generally have adopted positions consistent with business interests represented by the Chamber. In the early 1980s the Chamber had approximately 80 employees, a staff size larger than Houston's Planning Department. With respect to the political influence, one report noted that "one had only to study the goals listed by the Chamber of Commerce each year to get a good indication of what City Hall would be working on in years and decades to come" (cited in Feagin, 1988, p. 165).

The Chamber was renamed the Greater Houston Chamber of Commerce in 1987 and was restructured in 1989. That year, Houston's business leadership created the Greater Houston Partnership (GHP). The Partnership is an umbrella organization that coordinates the activities of four city advocacy groups. At the time of the restructuring, Lee Hogan, a former president of the Houston Economic Development Council (HEDC), said it "demonstrates the concept of one umbrella organization—the Greater Houston Partnership—overseeing Houston's business advocacy." Hogan stressed that the realigned structure permits "a focal point for setting public policy on business climate, business activity, job creation, economic development, and quality of life issues" (Greater Houston Partnership, 1989a, p. 1). Hogan believes the revamped structure will be more effective in lobbying for state assistance than efforts by scattered groups with divergent agendas.

According to the Partnership's leaders, the new organizational structure speaks for all of Houston's business community. Yet, opinion about the Greater Houston Partnership is divided within the elite economic community, not to mention within and between nonbusiness groups. Clearly, all of Houston's business leaders have not embraced the activities of the GHP and particularly the HEDC, the so-called public-private partnership created by the Houston Chamber.

THE HEDC: HOUSTON'S PRIVATE-PUBLIC PARTNERSHIP

In the 1970s, as the public perception grew that urban renewal programs were ineffective, many local governments in the United States experimented with partnerships with industry and other interest groups. Organizations specifically created to cultivate businesses seemed superfluous during Houston's long expansion. As a result, the city did not develop such a partnership until 1984, when the Houston Economic Development Council was created. By that time some 15,000 such agencies were operating nationwide.

Houston's public-private partnership is as unique among U.S. cities as is its business elites' ability to direct governmental affairs. In many cities, the public sector has taken the initiative in creating the partnership and then sought the support of the business community. That approach has been reversed in Houston.

The HEDC emanated from the private sector and acts without public sector interference, despite receiving a third of its budget from local government offices. In March 1986 the City Council voted unanimously to match private contributions to the HEDC. Mayor Whitmire initially resisted city funding, arguing it would mean higher taxes. The Greater Houston Political Action Committee, a business PAC, worked hard to get the matching grant passed. Members of the local business leadership called it the "first true economic wedding of the public and private sectors" (Hart, 1986, p. 1A).

At first glance, public funding for the privately created business organization appears to make sense. According to *Site Selection* magazine, the HEDC is among the top economic development groups in the country (cited in Greater Houston Partnership, 1989b, p. 1). In supporting this assessment, the magazine highlighted the 26 businesses that located or expanded in the Houston area in 1988. One important reason the magazine gave the HEDC rave reviews was its ability to persuade Compaq to stay and expand in Houston. Yet, this "success" story was greased by a multimillion dollar package of incentives, including training programs, recruiting assistance, tax breaks, and roadway expansions for which the residents of Houston will be paying a long time.

In the spring of 1989 the head of the HEDC, Lee Hogan, a former owner of a local engineering and construction firm, was promoted to president of the GHP. The top job at the HEDC was assigned to John Brock, who had been senior vice-president of marketing for Paine Webber Properties, Inc. At the time of the reorganization, Brock lauded the support given to the HEDC by the Houston business community: "I believe that the tremendous support we received from the business community on various projects has been the major contributor to HEDC's success" (GHP, 1989c, p. 8). Brock's comments are significant and more revealing than he probably realized. The emphasis placed upon the business community's involvement in Houston's *private-public* partnership demonstrates how little grassroots groups have participated in the organization.

Since its inception, the HEDC has attracted criticism about its goals, methods, salary structure, and booster programs (such as Houston Proud and Houston Beautiful) that many observers see as lacking in substance. But the organization pushes ahead, unrelentingly optimistic. Exaggerating the current modest recovery, the economic development group recently changed its slogan to Houston, Back on Top to Stay.

RACIAL POLITICS

Racial politics has affected this free-enterprise city's past and present. Fred Hofheinz, son of Mayor Roy Hofheinz and local lawyer, served as mayor from 1974 to 1978 after his effort to enfranchise voters in the minority wards was rewarded with victory and with a more representative City Council.

Today the 4th Ward is filled with predominantly black renters who typically pay $150 a month to live in decaying, wood-frame, rat-infested houses. Ironically, the first free blacks who moved to Houston 120 years ago settled in this district, which was dubbed Freedmanstown. Italian families began to buy property there in the early 1900s, and in the 1950s the community was a center of small businesses, as well as a cultural hub. The area began to deteriorate in the 1960s when many landlords began moving to the suburbs and renting their houses to city residents. New construction in the area was prevented by inadequate sewer and water systems. According to Harris County tax records, the ownership of the 4th Ward consists of several Italian families who formerly lived there but who have moved to better neighborhoods, along with other absentee landlords, government entities, and a handful of owner-residents.

In 1986, after years of deliberation, 225 of the landowners controlling 85% of the 4th Ward's property agreed to sell their holdings in a group transaction. The few owner-occupied residential dwellings left belong largely to African-American families who have lived in the neighborhood for generations. These families generally lack the money to invest in houses in another neighborhood and fear that the current redevelopment plan will give them "slum pay" for their homes—not enough to buy in another neighborhood. Many renters also fear that redevelopment would force them out of the neighborhood.

A current locus of Houston's racial politics is the Allen Parkway Village, a 1,000-unit HUD housing project in the 4th Ward. The public-housing project currently is facing its fourth demolition plan. Many residents believe that city officials simply need to increase the quantity and quality of existing services to make the housing adequate. This fourth redevelopment proposal involves the purchase and destruction of hundreds of buildings in the area just west of downtown Houston. Not only would the plan displace low-income residential property with office buildings and luxury condominiums, but also it is seen in symbolic terms by developers as the pivotal obstacle in the way of a more grandiose redevelopment effort in the 4th Ward.

A coalition of city church leaders has expressed grave misgivings about the latest development plan. A 4th-Ward clergy leader, Rev. F. N. Williams, commented: "We're not interested in the present residents being shipped out

God-knows-where so that they can make this another elite district. If what American General, Cullen, [U. S. Congressman Craig] Washington and the city are really trying to do is drive the present residents into the ground, then what is the difference between that and what happened to the Indians? I guess we're supposed to find another reservation somewhere" (L. Rodriguez, 1990, p. 16A).

Another group opposed to the elites' version of redevelopment is the Freedman's Town Association, a community-based, nonprofit organization. Gladys House, the group's founder, wants to oversee the new development of the area from a "grassroots" or low-to-moderate-income standpoint and to return Freedman's Town to an era when it was thriving with black homeownership and black businesses. House and her followers also would like to see the area declared a black historical district. Yet, American General Corporation, Cullen Center, Inc., and allied developers, whose development vision is fixed upon valuable property just west of downtown, continue to chip away at this grassroots political resistance. After protest ended the third demolition plan in 1989, the mayor and City Council stopped considering demolition schemes, but with Congressman Leland's death and yet another proposal, the political climate has changed. Houston's administration again appears open to development plans. The fourth proposal, supported by Leland's successor, Craig Washington, differs little from earlier versions. Issued in early 1990, the project calls for the demolition of all but 150 of the project's units. It also provides for the construction of 650 public housing units throughout the city, which would scatter widely the members of Houston's oldest black community.

Houston's recent race relations have been punctuated with periodic out-bursts of tension and conflict. After Congressman Leland died in a plane crash in 1989, the City Council met to consider renaming the airport in his honor. During a meeting, 16-year City Council veteran Jim Westmoreland, who opposed renaming the airport, said, "The idea now is to name the airport Nigger International. That way it would satisfy all the blacks. The other idea is that we paint all those white crosses [drunk driving markers] black." Several black community leaders, mayoral contender Fred Hofheinz, and the NAACP called for the at-large councilman's resignation. Even a Harris County commissioner who opposed the resignation idea called the comments "inexcusable," "obviously bigoted and stupid." Westmoreland, who has joked publicly about his high level of absenteeism at City Council meetings, offered an apology but no resignation (Jones & Gravois, 1989, p. A1).

Organized opposition from working class and nonwhite groups has seemed a remote possibility for many years, but enfranchisement of minority groups was a significant step and has altered the political atmosphere. Fourteen years after the 1944 U.S. Supreme Court decision (Smith v.

Allwright) that eliminated the all-white Democratic primary, Houston elected its first black to a public office when Hattie Mae White was elected to the Houston Independent School District (HISD) board. Six years later, another black—Asberry Butler—was elected to the same board. While two blacks were elected to state offices from Houston just one year after passage of the Federal Voting Rights Act of 1965, it was not until 1971 that Houston elected its first black to the City Council.

Houston's minority electorate grew more active in the 1970s. By the end of that decade, nonwhite voters had their strength diluted by at-large Council elections and the city's annexation of predominantly white suburban areas. In 1979 the U.S. Justice Department ruled that the annexations violated the Voting Rights Act. As a result, the Houston City Council grew from an 8-member at-large system to the current 14-member Council, of which 9 members are selected from districts and 5 are elected at-large. The first election held with the revamped structure saw two blacks elected from districts, a third black elected at-large, and the first Hispanic elected to the Council.

The likelihood that civil rights will become a salient issue in the near future appears to have diminished recently. Some of the city's most prestigious black political leaders have died, and there seems to be a dearth of prominent African-American advocates to carry on the political struggle. In addition to Congressman Leland, two other black civil rights pioneers in Houston died recently: Moses LeRoy and Christia Adair. LeRoy, a lifelong champion of minority issues, filed the suit in 1975 that resulted in the creation of single-member districts for the Houston City Council.

Clearly, the most important change in voting behavior in Houston between 1960 and 1990 centers on growing minority political participation. With the enfranchisement of many African Americans and Latinos, politicians have been forced to listen to the concerns of minorities. The administrative switch to single-member districts has meant greater representation of nonwhites in Houston's local government and a greater impact of minorities. But their influence is not in the key area of economic development. Minorities are making themselves felt in terms of social issues and having a voice in City Hall, but the economic course of the city has been largely insulated from the political pressures of minority groups.

HOUSTON'S POLITICAL REGIME

Historically, a coalition of Houston's economic leaders has been involved in the selection of the city's political leaders and in serving in political office. Houston's business elites traditionally have emerged out of the dominant economic engine of the period. In the 1830s and 1840s, agricultural production was the main economic force, with cotton and sugar cane constituting the most significant crops. Between the 1880s and the 1910s Houston became an important regional center for railroads and banking. Between the 1920s and the 1960s, primary commodity production became the key emerging economic sector, including timber harvesting and the extraction of sulfur, salt, lime, and other minerals. In 1901 oil was discovered 90 miles east of Houston; subsequent discoveries closer to the city through 1919 made Houston a major oil-and-gas exploration and production center in coming decades, becoming the major U.S. center for petrochemicals in the 1930s and the 1940s. The last 30 years have witnessed Houston's ascendancy as a national leader in medical and space technology. Each of these major economic transitions brought changes in the business elites, the longest lasting being the "Suite 8F crowd," which was influential between the 1930s and the 1960s. It is crucial to note that almost all of the sustained coalition building and formation has been within fractions of the business community.

Gradually, Houston's political structure has evolved toward a strong-mayor form, with the mayor increasingly gaining control over municipal department heads. A succession of business-oriented mayors has benefited from the strong-mayor type of government.

In the years between 1960 and 1990, Houston's city government has continued to revolve around a strong mayor. All department heads, with one important exception, are appointed by and report directly to the mayor. The city controller is the only other elected city executive and as such has become the second most powerful elected official in the city. The strong-mayor structure has been conducive to the domination of the local government by business elites. Before the mayor meets formally with department heads and the City Council about major projects, he or she has often already met with business leaders. Since the mayor typically controls the agenda, preference generally is given to business projects.

Between the late 1940s and the early 1970s, Houston had but four mayors. The four—Oscar Holcombe, Lewis Cuttrer, Roy Hofheinz, and Louie Welch—were all dependent upon the "Suite 8F crowd." Much evidence suggests that the core members of the business elite in Houston historically have had control over the major activities of the mayor's office. Between 1960 and 1990 all of Houston's mayors have either emerged from the business elites or have been business oriented.

Political change in Houston, in contrast with many other cities in this volume, has been a matter of degree rather than of kind. Changes have occurred in the politics of this big city since 1960, but they have not been fundamental changes. For example, working-class participation in the economic structure of the city has remained unaltered, and the city's racial relations have not fundamentally changed. In essence, the political formula in use 30 years ago is the same as that currently in place. From 1960 to 1990 all of Houston's mayors have been a part of, or have been primarily sympathetic to, the business class.

Nonetheless, in the 1970s and 1980s several business-oriented mayors with relatively liberal *social* views have been elected to office. Particularly noteworthy is Kathy Whitmire, who has been mayor since 1982, and Fred Hofheinz, who was mayor between 1974 and 1978. Both politicians promoted progressive causes, perhaps because both emerged from the moderate faction of the city's business elites.

A moderately progressive mayor, Fred Hofheinz benefited electorally from Houston's newly enfranchised minority voters. Once in office, Hofheinz hired high-level executive minorities for the first time. Another notable departure by Hofheinz was the decision to accept federal aid—for social and community groups—that had been routinely rejected by earlier administrations. Hofheinz also expanded the function of planning in governing the city, creating divisions of economic development and community development. Hofheinz located these divisions, traditionally functions of a city planning department, within his office. Perhaps of greater interest in terms of political change in Houston is the electoral victory of Kathy Whitmire in 1982. Her administration altered the relationship between local city government and the business elites. Several business leaders were leery of Whitmire for her progressive position on feminism and gay rights. Yet, she came to the mayor's post as an accountant, was conservative on most economic and political issues, and was strongly probusiness. As long as the city retained the progrowth and low-tax philosophy, Whitmire fit in neatly with the succession of business-oriented mayors, and she has been supported, if not always enthusiastically, by Houston's business leadership.

Like other local governments with a progressive-reformist orientation, Whitmire's administration emphasized efficiency. She replaced the letting of contracts through the "good ol' boy" network with a competitive system. She hired professional administrators where mediocre department heads had earlier reigned. She hired an African American to head a predominantly white police department notorious for its brutality and overt racism. Moreover, her 1982-1983 city budget was the first in 25 years to be presented and adopted on time. Some of her decisions, such as not to support the Chamber-created Houston Economic Development Council (HEDC), led to an active campaign in 1985 that pitted Whitmire against former mayor Louie Welch. Despite the opposition mobilized against her, Whitmire remained committed to her socially progressive yet probusiness approach to running local government and remained mayor.

In the context of political change, it is important to remember that Whitmire did make token changes that reflected important strides in the direction of women's rights, minority rights, and more neighborhood participation at City Hall. Yet, when Whitmire faced a serious fiscal crisis in 1987, she adopted the traditional conservative posture and pledged no new taxes. Instead, she laid off more than 300 city employees and raised a number of fees for city services.

Houston's most recent mayoral campaign, featuring Whitmire against Fred Hofheinz, showed that Whitmire has moved ever closer to the business elites and their policies. Richard Murray, professor of political science at the University of Houston, asserts that Whitmire began as an outsider but has become very much an insider over time. She may have come to the job as an "outsider," but she was still an accountant and businessperson. In earlier years, Whitmire's and Hofheinz's reformist reputations figured prominently in their mayoral campaigns, with both taking progressive positions on homosexual, minority, and abortion rights. While both politicians continue to advocate such causes, they are now accorded low priority, compared with issues surrounding the recent economic bust and how best to recover from it.

In short, the 1989 campaign featured both of Houston's one-time progressive candidates posturing as economic-development leaders and defenders of the good business climate. During the campaign, Whitmire never deviated from the business elites' position, championing the privatization of city services and vowing neither to raise taxes nor to institute zoning ordinances. She also seemed unconcerned with organized labor's opposition and is reported to have proudly proclaimed that "it is no secret that the unions are mad at me. They have reason to be mad at me" (Suro, 1989, p. 6). Hofheinz toed the same line and went on the offensive, accusing Whitmire of letting Dallas take the lead in luring corporations to the state. Hofheinz's hopes for victory were hampered by the untimely demise of Representative Mickey Leland. Leland was widely regarded as the city's most prominent black politician and was one of many community leaders that Whitmire had angered over the years. Before his death, defeating the mayor had become a personal issue with Leland.

Whitmire, while giving lip service to progressive causes, often fails to follow through. For example, according to the Texas Ozone Task Force, nearly every city in the country with an air pollution problem as bad as Houston's has argued for a tougher bill. But Whitmire, ever protective of the good business climate, states she does not want a law that could punish the city in the event industries do not have the technology to meet annual emission-reduction requirements. In response to more environmentally minded groups that favor clean-air mandates, the mayor refers to sanctions as "calamities."

The most significant change in electoral patterns between 1960 and 1990 has been the inclusion and growing participation of minority voters. Through the 1940s and 1950s the all-white Democratic primary and the Texas poll tax significantly reduced the black vote. The passage of the Voting Rights Act of 1965 resulted in growing numbers of black and Hispanic voters by the early 1970s. In part, the expansion of voter power stems from structural changes forced upon the white establishment through a suit undertaken by members of the city's minority communities, the suit that forced the replacement of an eight-member at-large City Council with one made up of nine members elected from specific districts and five elected at-large. Feagin has argued that "This was perhaps the most significant internal change in Houston's politics since the so-called 'progressivism' brought to City Hall by the business elite in the early 1900s (Feagin, 1988, p. 252).

The voting power of minority voters brought partially desegregated city councils, as well as white mayors such as Fred Hofheinz and Kathy Whitmire, who were more sensitive to their needs. Previously, the business elites and the local government were able simply to ignore inner-city communities; with several minority members of the community on the City Council and in other city offices, important services gradually have been provided to non-white constituencies.

EXTRATERRITORIAL JURISDICTION

As with many metropolitan areas, the Houston region is crosscut and interconnected with many local governments, including Harris County's hospital and flood control districts. The city also is surrounded by approximately 400 special utility districts, usually known as municipal utility districts (MUDs), that serve small subdivisions or other private developments in unincorporated areas. In addition, Houston interacts with 32 other cities located in or partly in Harris County.

Houston's physical dominance is made possible by a state law that provides Texas cities with an extraterritorial jurisdiction (ETJ). Houston's ETJ extends 5 miles from its corporate limits, covering in excess of 2,000 square miles. It includes the entirety of Harris County and parts of six adjacent counties. Houston's ETJ is a very powerful land-use instrument, giving the city exclusive authority to annex areas within its jurisdiction without voter approval, to prevent new incorporations, and to permit the creation of special utility districts. Despite the rights that Houston enjoys within its ETJ, it has no responsibility to provide services until it formally annexes an area.

Other cities in the region have ETJs, but none exceeds Houston's, and the two largest cities—Pasadena (112,560) and Baytown (56,923)—are constricted by their geography. Like other cities to the east and southeast of

Houston, Pasadena and Baytown are constrained by Houston's ETJ, into which they cannot expand, and by the Gulf Coast. Another 13 cities in Harris County are communities within Houston and do not have ETJs.

Demographic changes in Harris County between 1970 and 1980 demonstrate Houston's control over other Harris County cities. During that time, Houston and its ETJ captured 89.6% of the population growth, while the remaining 32 cities in the area received just 10.4% of the total. Houston restricts the population growth of other cities by constraining their ability to expand their borders. Meanwhile, Houston has significant potential for expansion both within its corporate limits and its ETJ.

Because such a high proportion of the population growth in southeast Texas has been within Houston's political boundaries, the suburbia-central city conflict that one finds in many other metropolitan areas has been minimal. Instead, at least until the 1970s, suburban politics was transformed into an internal struggle between business leaders within the city. In effect, Houston's business leaders and elite economic actors from outlying residential areas ran the city. Since the 1970s the rise of minority voters has added more political struggles that were organized, at least in part, with a recognition of minority concerns. In this way, the white within-the-city voters have provided a "suburban" force against the rising political power of minorities. A key issue for those outside the city boundaries has been the racial composition of Houston's school system which, since it desegregated in the late 1970s, has fostered some suburbanization.

Houston's major suburbanization took place from the late 1940s to the early 1980s. A construction boom was a part of the city's robust economy in the 1950s, as roughly 100,000 new homes were built, significantly expanding suburbia in the metropolitan area. Much of the residential development in recent decades has been in large-scale tract developments farther out from the central city. By the 1970s, the outlying area had become a sprawling complex of suburban subdivisions. In one 12-year period ending in the late 1970s, Houston developers built several hundred suburban subdivisions. Subsidized highways and housing made extensive suburbanization possible, and Houston's lack of zoning ordinances promoted the proliferation of suburbs along the interstate highway spokes radiating from Houston's downtown.

Annexations by Houston since World War II are key to understanding the city's suburban development. Annexation has been used routinely by the city to increase the tax base and to capture suburban growth. Houston's annexation policy was used most extensively between 1949 and 1956, with more modest expansions between 1960 and 1990. Since 1948 the city's size has increased by a factor of 7, from 83.7 square miles to 556.4 square miles.

Today, Houston's territorial size is larger than Chicago and Philadelphia together, and its ETJ is larger than the size of Rhode Island.

CONCLUSION

Houston's business elites have successfully held sway over the political process between 1960 and 1990, as well as before then. As the 1960s began, the remnants of the "Suite 8F crowd" still had an important degree of influence over citywide decision making, but by the late 1960s and early 1970s the switch to a corporate-based business leadership was well underway. By 1990 the Greater Houston Partnership was a firmly entrenched structure with a planning and economic development staff surpassing that of the city.

Although for much of the past decade Houston's business elites have had to contend with an unprecedented economic decline, an incremental economic expansion was evident as the 1980s ended. Despite the modest economic growth, the current recovery is being popularized in typical Texan style by local business advocacy groups and ordinary Texans alike. For example, the local leadership downplays the extent to which the local economy depends upon government subsidies. In this way, the business elites have managed to have their cake and eat it too: a generous subsidy program for private projects, while maintaining the essence of the good business climate. Were it not for a steady flow of revenue from the federal government for the Johnson Space Center and federal support for Houston's ailing financial institutions, the local economy could still be mired in a recession.

Houston had a minimal number of mayors between 1960 and 1990. From the early 1960s to the mid-1970s, the most cohesive business elites in the city's history provided the candidates or strongly supported those who won. Beginning in the mid-1970s, an impact for previously disenfranchised voters began to materialize. Yet, even the two most "progressive" of the candidates to benefit from the African-American and Hispanic vote are strict fiscal conservatives and protectors of Houston's cherished "good business climate." The rhetoric (and some actions) flowing from City Hall may have moved in a progressive direction on social issues, but the economic direction shaped by the city's business leadership has not wavered since 1960.

As of the late 1980s, significant challenges to the business dominance of Houston appear remote, with minority influence seeming to have reached its zenith, particularly given the recent loss of key minority leaders. As Shelton et al. summarize, "For the foreseeable future, it appears that the business of Houston's local government will continue to be business" (1989, p. 136).

REFERENCES

Belkin, L. (1989, August 22). Now, it's remember the oil bust! *The New York Times*, pp. 25, 44.
Davidson, C. (1982). Houston: The city where the business of government is business. In W. Bedicheck & N. Tannhill, (Eds.). *Public Policy in Texas* (pp. 275-288). Glenview, IL: Scott Foresman.
Feagin, J. R. (1988). *Free enterprise city: Houston in political-economic perspective*. London: Rutgers University Press.
Graczyk, M. (1988, July 5). Tarnished Houston regaining some of its Sunbelt luster. *San Antonio Light*, p. 1D.
Greater Houston Chamber of Commerce. (1989a, January). *Houston economic overview* (Prosperity 1974-1981), p. 3.
Greater Houston Partnership. (1989a, June). Partnership organization complete. *At Work*, p. 1.
Greater Houston Partnership. (1989b, June). HEDC named one of the nation's best. *At Work*, p. 1.
Greater Houston Partnership. (1989c, June). An interview with the new HEDC president reveals insight, plans. *At Work*, p. 8.
Hart, J. (1986, March 24). Council wins fight to fund HEDC. *Houston Business Journal*, p. 1A.
Houston Economic Development Council. (1989, March). *Houston: Back on top to stay*. Houston: Author.
Jones, R., & Gravois, J. (1989, October 28). Black leaders blast Westmoreland. *Houston Post*, pp. A1, A14.
Port of Houston Authority. (1989, December). *Port of Houston*. Houston: Author.
Rodriguez, L. (1990, February). Legacy in jeopardy, *Houston Chronicle*, pp. 1A, 16A.
Rodriguez, N. P. (1990, March 4). Foolish if Houston ignores its new faces. *Houston Chronicle*, p. 1F, 4F.
Shelton, B. A., Rodriguez, N. P., Feagin, J. R., Bullard, R. D., & Thomas, R. D. (1989). *Houston: Growth and decline in a Sunbelt boomtown*. Philadelphia: Temple University Press.
Smith v. Allwright (1944).
Suro, R. (1989, October 30). Houston candidates court business. *The New York Times*, p. 6.

12

Los Angeles: Transformation of a Governing Coalition

ALAN L. SALTZSTEIN
RAPHAEL J. SONENSHEIN

THIRTY YEARS AGO, Edward Banfield described Los Angeles city government as "pre-Civil War." The mayor, he argued, was "almost too weak to cut ribbons" (Banfield, 1963, p. 80), and the major functions of government were divided between a strong City Council, numerous autonomous departments, and an electorate with considerable statutory authority through initiatives and referenda. He found power widely dispersed and central authority nearly nonexistent. Banfield and co-author James Q. Wilson claimed that because of these traits, ". . . many things cannot be done because it is impossible to secure the collaboration of all those whose collaboration is needed" (Banfield & Wilson, 1963, p. 111).

There is evidence that within the past decade Los Angeles has undergone economic, social, and cultural upheavals deeper and broader than those experienced by any other large American city during the same period (Lockwood & Leinberger, 1988). The metropolitan area is fast becoming the hub of the Pacific Rim and may be rivaling New York for preeminence in finance and production. It has become the port of entry for the largest number of foreign-born people in the country. Its several decentralized urban centers are thought to be the model for city plans of the future. New cultural facilities and expanded universities are promoting an international identity for the urban area.

Could such changes occur within the kind of political system Banfield and Wilson (1963) described? Behind the social, cultural, and economic upheavals that underlie the new Los Angeles were an equally important series of political changes. New coalitions of minority groups, business leaders, citizens' groups, and the federal government assisted and charted the emergence of Los Angeles as a global city. The political system of present-day Los Angeles, while structurally similar to that of the 1950s, has developed informal ways of increasing authority of elected officials. In turn these elected officials facilitated the transformation of the city. More recent changes in turn have undermined the stability of the political regime. This chapter grows out of research conducted by the authors over a period of several years. One research program has explored the impact of federal aid on the city's political structure, formal and informal (Saltzstein, Sonenshein, & Ostrow, 1986). A related study charted the increasing power of the Los Angeles City Council as the Bradley mayoral administration gained strength and then encountered difficulties (Saltzstein & Sonenshein, 1986). The research for those studies involved archival research, participant observation, and more than 20 elite interviews with elected and appointed officials in the city and with federal funding officials.

A second research program has explored minority political mobilization and biracial coalition politics in Los Angeles, particularly the development of the biracial coalition behind Tom Bradley (Sonenshein 1986, 1989, 1990). This research has sought to apply historical techniques to the study of a big-city biracial coalition, exploring its evolution over a 25-year period. The project has involved more than 25 elite interviews, substantial archival research, voting analysis, and participant observation.

This chapter joins the two programs in exploring both electoral and governmental developments in the 1960-1990 period. We begin by describing the changes that have occurred economically and socially in the Los Angeles region from 1960 to the present day. Next, we evaluate city politics in Los Angeles during the same period, including the emergence of a liberal coalition under the leadership of Mayor Tom Bradley. Last, we discuss the changes in city life in the 1980s that have led to a decline in the influence of the ruling coalition.

A MULTICULTURAL WORLD CITY

The Los Angeles we refer to is a city embedded within a county and a region. The region is usually defined as the five counties that stretch from the Arizona border to the ocean on the east and Ventura to Orange counties north to south. This complex is arguably several urban areas, but to some degree all parts of it are interrelated economically and socially. The Southern California Association of Governments (SCAG), an advisory council of governments, provides some guidance for regional policies, and other regional units comprise all or parts of this area.

Los Angeles County contains a 1990 population of 8.7 million people. The city of Los Angeles, with 3.5 million residents in 1990, is the dominant government of the county. It is the second-largest city in the nation and one of the only big cities to register a large population gain in the 1980s and 1990s (15.4% from 1970 to 1980, 17.2% from 1980 to 1990). Such major cities as Long Beach, Pasadena, and Glendale are separate municipalities, and over 1 million residents live in county unincorporated areas. Many county residents live near the Los Angeles city center. Thus, both city and county are important forums for decisions in the Los Angeles urban area.

The land area of the city of Los Angeles totals over 464 square miles, making it one of the physically largest cities in the country. The municipal boundary contains the traditional urban center, many areas of newer development and a considerable amount of open land. The "central city-suburb" distinction has little meaning in Los Angeles since the city contains areas similar to the traditional suburb, and several smaller cities are surrounded by the city of Los Angeles.

Los Angeles today is a city of contrasts. Economically the city has emerged as a "world metropolis" with a diverse industrial base and an important leadership position in the Pacific Rim. Some have argued, in fact, that the city will become the hub of a new commercial empire by linking the new economies of East Asia with parts of Latin America and the United States. Poverty and social deprivation, however, also have increased during these same years. A social transformation rivaling the early years of the Industrial Revolution on the Eastern Seaboard has occurred. The population has changed from largely Anglo to multiracial and multiethnic with a large foreign-born population.

Manufacturing jobs in the region have expanded significantly, accounting for over one fourth of the increase in manufacturing in the United States from 1970 to 1980 (Soja, Morales, & Wolff, 1986). Los Angeles also has emerged as the financial hub of the Western states, surpassing San Francisco in total deposits and savings over the past decade (Lockwood & Leinberger, 1988). The ports of Los Angeles and Long Beach now command the largest total tonnage in the country, surpassing the New York-New Jersey Harbor during the past year.

Banfield commented that in 1965 ". . . so many people now live in outlying sections and in suburbs that downtown is no longer important to politicians" (Banfield, 1965, p. 88). Indeed, the downtown of 1960 was rather small, drab, and lacking in the vitality normally associated with the center of a vibrant city.

By contrast, today's downtown is a lively, crowded center of commerce, industry, and culture. A building boom expanded office space in the downtown area by 50% from 1972 to 1982. While still only one of several "urban villages" within the urban area, recent years have seen downtown assert itself as the preeminent center of the metropolis.

The expansion of wealth and power, however, has not been shared by all residents. In fact, there are many indications of increased poverty, an expanded underclass, and greater differences in wealth. In 1960 the median income of both the county and the city were significantly higher than that of the state. In 1970 the median income of the three entities was very similar. By 1980 the median income of Los Angeles was more than $2,000 less than that of the state and over $1,500 less than the county. Thus, over time, the city has become relatively more deprived than the county and state.

The proportion of city families below the poverty line rose from 9.9% to 13.0% between 1970 and 1980. The county's percentage increased from 8.2% to 10.5% during the same period. Increases in unemployment also have occurred in recent years. The number of homeless people in the city is estimated conservatively in the tens of thousands. Thus, economic dislocations and inequities have increased considerably, resulting in demands for improved social services, better police protection, and economic and social aid.

The Los Angeles of today is one of the nation's most racially and ethnically diverse cities. Much of this diversity has developed over the past 20 years. In 1960 the combined black, Latino, and Asian population of city and county was about 18%. Today these groups are a majority of both entities (see Table 12.1).

Much of the Asian and Latino increase derives from new immigrant groups entering the city. The foreign-born in the city have increased from 9.5% in 1960 to 27.1% in 1980. The county's percentage changed from 9.5% to 22.3% in the same period. Both categories mask considerable diversity. The Asian category includes many Southeast Asian refugees, as well as significant numbers of Chinese, Koreans, and Filipinos. The Latino category includes numerous Central American refugees, as well as significant numbers from Mexico.

Los Angeles government has been influenced strongly by the reform tradition. While a separately elected mayor is the focus of city policy making, the power of the mayor's office is limited by numerous structural constraints. The 15-member City Council has considerable power over appointments and budgetary matters. Council members are nominally nonpartisan, with each district containing over 200,000 people. Several independent commissions led by appointed commissioners govern city departments. The Council must approve the mayor's commission appointments. Last, most personnel decisions are made by an independent Civil Service Commission. The mayor and Council have very few patronage positions at their disposal. All elections are nonpartisan, and national parties have not exercised significant influence on Los Angeles politics.

TABLE 12.1

| | Population | | Percent Black | |
	City (000)	County (000)	City	County
1960	2,479	6,039	13.5	7.6
1970	2,512	7,042	17.9	10.8
1980	2,900	7,478	17.0	12.6
1990	3,400	8,700	13.0	14.0

| | Percent Latino | | Percent Asian | |
	City	County	City	County
1960	*	*	1.6	0.8
1970	18.4	14.9	2.1	1.3
1980	27.5	27.6	6.9	8.9
1990	39.9	39.9	9.2	9.7

| | Percent Foreign Born | |
	City	County
1960	12.5	9.5
1970	14.6	11.3
1980	27.1	22.3

SOURCE: U.S. Census (1990 estimates).
*Latino counts were not included in the 1960 census.

Implicit in these structures is an assumed separation of policy and management and a very weak role for elected officials. Reformers assume that city problems can be managed best by technically trained managers and competent personnel. Only guidance of a very general sort is expected from elected officials.

The years from 1960 to the present, however, have witnessed increased pressures for more political direction. Much of the pressure came from a steadily growing coalition of minorities and white liberals alienated by the traditional low profile of the city's political leaders. A stronger mayor and City Council emerged to respond to the perceived need for more centrally directed policies.

A GOVERNING REGIME EMERGES

A ruling coalition of blacks and white liberals (especially Jews), some Latinos and Asians, downtown business and labor interests, and community-based organizations and the federal government developed in Los Angeles between 1960 and 1985. The coalition's success strengthened the power of elected officials under the leadership of Mayor Bradley. The coalition implemented a series of policies that responded to the social and economic changes occurring during this era. This section discusses and analyzes the development of this coalition.

Banfield's chapter (1963) ends with the election of Sam Yorty as Mayor in 1961 and the early stages of black electoral success. Yorty was an insurgent, populist candidate with an unlikely coalition of white suburban homeowners and minorities. He defeated Mayor Norris Poulson despite Poulson's support by downtown business interests, the traditional source of influence for Los Angeles mayors.

Yorty viewed the mayor's job as separate and distinct from the Council and proceeded to lock horns with key Council members on most major issues. Politics during these years revolved around conflicts between the mayor's office and the Council.

In 1963, three black City Council members were elected, suddenly increasing black representation from none to 20% of the Council. One member, Tom Bradley, was elected in the multiracial 10th Council District with the support of blacks, Jewish liberals, and Asian-Americans (Sonenshein, 1986). The reform wing of the Democratic Party (the California Democratic Club movement) played a central role in organizing the Bradley campaign.

The 1965 Watts riot was a watershed for the city's political life. The disorders divided the city along racial and ideological grounds. Mayor Sam Yorty became the embodiment of white conservative resistance to social change, including reluctance to pursue federal aid and downtown redevelopment. On the other side was an emerging citywide coalition led by Bradley and including the mobilized black community, white liberals (principally Jews), and a minority of Hispanics (Sonenshein, 1989).

During this period, conflict was persistent in City Hall between Yorty and Bradley over whether to pursue federal aid. Los Angeles received a low share of federal money, relative to need and population, and Yorty's dramatic testimony in congressional hearings highlighted his resistance to "federal interference" (Saltzstein et al., 1986).

Meanwhile, the liberal coalition in the Council was developing an agenda for social change that would not strain city revenues. This platform involved minority representation, civilian control of the Los Angeles Police Department, aggressive pursuit of federal aid, and downtown redevelopment. Bradley and his allies, for instance, created the Board of Grants Administration, over Yorty's veto, to search for new federal and state funding for city projects (Bruce, 1974).

Bradley emerged as a potential mayor during his years on the Council. His background and career prepared him for the unusual role he would play as a black mayor of a city with a relatively small black population (about 18% in 1973).

Bradley was born in 1917, one of seven children of Texas sharecroppers. The family moved to Los Angeles when he was 10. The future mayor excelled in the public schools and entered UCLA on a track scholarship (Robinson, 1976). He left the university in 1940 without his degree and joined the Los Angeles Police Department. He rose through the ranks despite rampant racial bias and eventually became the city's second black lieutenant. Bradley retired from the police department in 1961 to become a practicing attorney and to enter politics. His family also moved to a racially mixed neighborhood on Los Angeles's West Side, and he made many contacts within the reform wing of the Democratic Party.

Bradley challenged Yorty for the mayoralty in both 1969 and 1973. Yorty defeated Bradley in 1969 with the use of a brutal, racially polarizing campaign, but in 1973 Bradley prevailed in a less racially charged time. Now the liberal coalition was in a position to implement its platform of political, social, and economic change.

As a governing coalition, the Bradley administration began as a linkage between black and white liberals and the emerging environmental movement. The regime also was closely tied to national Democratic leaders and federal funding officials; Bradley provided a vehicle by which money could be poured into the largest Western metropolis. With federal funds Bradley was able to expand the power of elected officials. Of course, this particularly strengthened the mayor's office. Even the City Council gained power as federal funds were temporarily used to hire additional staff for Council members (Saltzstein & Sonenshein, 1986).

Federal funds, especially during the Carter years, were used in Los Angeles both to expand social services and to pursue economic development. The administration used federal economic development grants to save such endangered job-producing industries as the tuna cannery in San Pedro and the flower market in downtown Los Angeles. Substantial federal funds also were expended in job-training programs that in turn built links to neighborhood-based organizations sophisticated enough to utilize federal funds.

The administration, in close alliance with allies in the City Council, quietly pursued other important elements of the liberal agenda. For instance, affirmative action in city hiring was augmented significantly and for the first time a city regime succeeded in placing some civilian control over the police department (Sonenshein, 1990). All three areas have been cited as significant goals for minority-liberal regimes (Browning, Marshall, & Tabb, 1984).

In 1975 the Bradley administration embarked on one of the nation's most ambitious downtown redevelopment programs. The Community Redevelopment Agency (CRA) employed tax increment financing and various debt financing mechanisms to rebuild downtown. Redevelopment required the destruction of low-income properties, replacing them with upper-class condominiums and office buildings.

Controversy surrounded the redevelopment issue, with conservative Council members opposing the plan. When the redevelopment program came to a vote, a wide range of liberal and minority spokespersons addressed the Council in favor of it. The ultimate agreement reached between the city and the county (which feared the loss of tax revenue) involved a downtown "spending cap" of $5 billion over the life of the redevelopment program—a limit that became an important issue in the late 1980s when redevelopment became considerably less popular among liberal constituencies.

Redevelopment helped facilitate a fundamental change in downtown Los Angeles, sustaining the construction of skyscrapers, new housing, and new shops and cultural centers. Los Angeles was becoming a "world class city" with major cultural and social aspirations. The Bradley regime was much identified with the transition of the city into a more cosmopolitan metropolis, a trend reemphasized by the successful 1984 Olympics.

At the same time, the redevelopment program changed the nature of the Bradley coalition from a minority-liberal political alliance to a much broader regime encompassing major business and labor interests. Campaign contributions from business and labor became major factors in the ability of the mayor and the City Council majority to maintain strong incumbencies and to preempt potential electoral challenges (Sonenshein, 1986). Further, the confinement of redevelopment to downtown helped allay possible environmental concerns in other areas of the city, particularly on the coastline.

In many ways the 1970s were extraordinarily stable for the city's political life, allowing economic shifts to continue untroubled. Abundant federal aid and tax revenue generated by redevelopment contributed to the administration's ability to hold together its coalition and to distribute political and economic benefits.

Bradley's personality and operating style allowed benefits to be shared without creating major racial and ethnic conflict. In one sense he dealt with dilemmas by limiting the number of conflict-laden choices. Bradley tends to operate behind the scenes. He rarely takes the lead on controversial issues and seeks instead to create quiet consensus among elected officials, bureaucrats, private interests, and community groups (Sonenshein, 1986). He also has managed to be nearly invisible on highly emotional issues, such as school busing, while at the same time mediating conflict between blacks and Jews.

In the process he avoids becoming the target of protests and willingly gives others credit for what may rightfully be his accomplishments. He avoids redistributive decisions, preferring to expand the city's resource base and letting City Council members haggle over distribution (Saltzstein et al., 1986).

Bradley's style has facilitated the resolution of many major problems with little public conflict. During the first 12 Bradley years there were few obvious losers. The City Council became a rather quiet, contented body. A clear majority of the Council developed close working relationships with the mayor, and he generally deferred to individual Council members on spending decisions in their districts. The Council had very little turnover during the early Bradley years.

Thus, the Bradley regime constructed a governing coalition consisting of liberals, most particularly blacks and Jews, along with significant elements of business and labor and powerful community-based organizations in minority areas. Resources from the federal government, and redevelopment and policy commitments from City Hall satisfied coalition members and expanded its size. Maintenance of this smooth-running machine, however, required expanding resources and acceptance of a growth-oriented philosophy.

STRAINS IN THE GOVERNING COALITION

By the late 1970s and early 1980s some pillars of the status quo shifted, and the regime found the ground below it less stable. Proposition 13 in 1978 and the election of Ronald Reagan in 1980 drastically reduced the levels of federal and state funding. The decline of external funds also meant fewer economic benefits at the disposal of elected officials, reducing the administration's strong fiscal ties to community organizations. The recession of the 1980s added to the city's difficulties. When poverty, homelessness, and unemployment increased, the city had fewer resources to respond. These changing conditions had a major impact on the decision-making process in the city.

The emergence of newly assertive ethnic groups, particularly Latinos and Asian-Americans, further strained the coalition. The 1982 Council reapportionment highlighted the continued absence of either group in elected offices, leading to a lawsuit against the city by Latino organizations (Regalado, 1988). Divisions within the core of the coalition (blacks, white liberals, and business/labor) also soon emerged over the city's "quality of life."

As economic growth spilled over from downtown into the affluent liberal Westside, these areas became clogged with traffic and pollution. A "slow growth" movement targeted the Bradley administration's close ties with downtown business as symbolized by Bradley's decision in 1985 to authorize oil drilling in the Pacific Palisades. At the same time, minority communities were becoming restive at the decline of social services. A shortage of low-income housing had been exacerbated by the administration's poorly run housing programs, and further criticism involved the increasing number of homeless people on downtown streets.

Little changed in the City Council from 1973 to 1985, and in only one case was an incumbent defeated at the polls. From 1985 to 1989, however, a series of electoral changes and new appointments created a Council of much greater diversity and independence. Contested elections occurred in several districts, and new issues were raised repeatedly. Five new faces were added to this Council of 15.

In 1985 Richard Alatorre, a powerful and ambitious state legislator, became the first Latino member of the Council since 1963, replacing the retiring Arthur Snyder in the 14th District. Though a supporter of Bradley, Alatorre maintained ties to Democratic Party leadership at the state level and acquired an independent power base in the Latino community.

Michael Woo became the first challenger to win a Council seat since 1977 when he defeated Peggy Stevenson in the 13th District in 1985. Woo was also the city's first Asian-American City Council member.

Court-ordered redistricting created a new Latino 1st District in East Los Angeles, where Gloria Molina, a state legislator and rival of Alatorre, became the second Hispanic Council member. Joel Wachs, a longtime supporter of the mayor, was redistricted into a more conservative 2nd District. He was reelected by carving out new policy positions more independent of the mayor.

The elections of 1987 dramatized the change in the mayor's influence. Longtime Bradley ally David Cunningham retired from the 10th District seat. Bradley, who had represented the district for 10 years before becoming mayor, campaigned vigorously for longtime friend Homer Broome. Broome was defeated nearly 2 to 1 by newcomer Nate Holden in the general election.

Pat Russell had represented the city's middle class in the 6th District since 1969. She had been Council president since 1983 and in that position coordinated the Bradley legislative agenda. The mayor also campaigned actively on her behalf, but she was defeated easily by Ruth Galanter, a planning consultant and environmental activist with little previous governmental experience. Several neighborhood groups loosely aligned with a burgeoning environmental movement were her main sources of support. Russell was replaced as Council president by conservative Richard Ferraro, Bradley's 1985 opponent for mayor.

The 1987 defeats of Bradley supporters in two contrasting districts revealed significant difficulties with two of the primary components of his coalition. The loss in the 10th District suggested that Bradley's influence with black voters and the lower middle class had diminished. The 6th District loss generally was attributed to the rising environmental movement and the perception of Bradley as a proponent of growth and development.

Bradley also faced frequent criticism in the press over alleged violences of proper ethical conduct. He had received fees from banks and savings and loans that had done business with the city. A city investigation cleared the mayor, but a federal probe continues. Though as yet no charges have been filed against the mayor and he has been cleared of many of the accusations, the Bradley image was seriously tarnished.

The 1989 race for mayor dramatized the influence of all of these changes on the mayor's electoral influence. Holden became a last-minute challenger when liberal Councilor Zev Yaroslavky, a visible City Council member and longtime perceived heir apparent, decided just three months before the primary not to run. Low in campaign funds and virtually unknown outside his home district, Holden was assumed to be a token challenger to Bradley. The mayor, however, won a narrow 51% majority in the primary. This total contrasts with previous winning margins of 59% in 1977, 64% in 1981, and 68% in 1985.

Bradley's actions since the 1989 election suggest the development of a new strategy to respond to the new forces in the political environment. The mayor's office has tried to shift its own focus away from a preoccupation with business, relinking itself to environmental forces. The mayor has appointed a majority of environmentalists to the Department of Water and Power (DWP). Previously, the DWP had been an apolitical, business-dominated agency with a commercial orientation. Bradley's new majority had orders to change the DWP into a more responsive energy agency. He also encouraged a refocusing of CRA funding to develop an innovative after-school program and new low-income housing projects.

In something of a race against time, Bradley and his allies have been trying to reconstitute themselves into the path of "urban populism" noted by De Leon and Powell (1989) in San Francisco, rejoining environmental and minority interests. Bradley's competitors are trying to achieve the same synthesis. The trend in Los Angeles is unmistakable—political leaders are under considerable pressure to find local resources and ideas to make the city work better. A growth agenda alone will not suffice.

The ability of the mayor or any other elected official to orchestrate this coalition, however, is considerably diminished. Bradley's influence over the

Council has declined. The loss in federal funds and state restrictions on the city's ability to raise taxes have limited severely the amount of revenue available. Inter-ethnic conflict may increase as Latinos move into black neighborhoods. Consensus building that is essential to political influence over the future transformations of the city is on perilous ground.

AN UNCERTAIN FUTURE

Los Angeles faces the 1990s with much uncertainty. The city prospered politically in the 1970s and early 1980s as a powerful coalition was able to harness new resources to satisfy the city's various interests. While minorities and white liberals still have the potential to form a ruling coalition, the situation is becoming more complicated. The early biracial coalition is strained.

New groups have emerged during a time when resources have declined and perceived problems have increased. Putting a new coalition together today would be more difficult than it was earlier, as the means to reward potential members are significantly fewer.

We must, therefore, return to at least some of Banfield's insights. Structurally, the city of Los Angeles discourages central leadership. The City Council and autonomous commissions are legally powerful entities. The mayor must garner resources from external sources to gain influence over the system. Bradley was able to do so largely through the use of federal funds and downtown redevelopment. With these two avenues closed off, the ability of any mayor to centralize power in the city is problematical.

Los Angeles is entering a new, uncharted era. The power of elected officials is likely to remain high but increasingly fragmented. The sophisticated city of the 1990s looks far different from what Banfield saw, though it contains the same political structure. Political struggles are likely to increase over the social and economic consequences of growth with unpredictable impacts on local politics.

REFERENCES

Banfield, E. (1965). *Big city politics.* New York: Random House.

Banfield, E., & Wilson, J. Q. (1963). *City politics.* Cambridge, MA: Harvard University Press.

Browning, R. P., Marshall, D. R., & Tabb, D. (1984). *Protest is not enough: The struggle of blacks and Hispanics for equality in city politics.* Berkeley: University of California Press.

Bruce, E. (1974, September). The grantsmanship game in Los Angeles. *Nation's Cities,* pp. 15-16.

DeLeon, R., & Powell, S. S. (1989, June). Growth control and electoral politics: The triumph of urban populism in San Francisco. *Western Political Quarterly, 42,* 307-332.

Lockwood, C., & Leinberger, C. B. (1988, January). Los Angeles comes of age. *Atlantic Monthly,* pp. 43-52.

Regalado, J. (1988). Latino representation in Los Angeles. In R. E. Villareal, N. G. Hernandez, & D. Neighbor (Eds.), *Latino empowerment: Progress, problems and prospects* (pp. 91-104). New York: Greenwood.

Robinson, J. L. (1976). *Tom Bradley: Los Angeles' first black mayor.* Unpublished doctoral dissertation, University of California, Los Angeles.

Saltzstein, A., & Sonenshein, R. (1986). *The city council resurgent? The case of Los Angeles.* Paper presented at the annual meeting of the Western Political Science Association, Eugene, Oregon.

Saltzstein, A., Sonenshein, R., & Ostrow, I. (1986). Federal aid to the city of Los Angeles: Implementing a more centralized local political system. In T. Clark (Ed.), *Research in urban policy: Vol. 2* (pp. 55-76). Greenwich, CT: JAI.

Soja, E., Morales, R., & Wolff, G. (1986). Los Angeles restructures into a world city. *Architecture and Planning* (pp. 15-18).UCLA Graduate School of Architecture and Urban Planning.

Sonenshein, R. J. (1986, Summer). Biracial coalition politics in Los Angeles. *PS 19,* 582-590.

Sonenshein, R. J. (1989, June). The dynamics of biracial coalitions: Crossover politics in Los Angeles. *Western Political Quarterly, 42,* 333-353.

Sonenshein, R. J. (1990). Biracial coalition politics in Los Angeles. In R. Browning, D. R. Marshall, & D. Tabb (Eds.), *Racial politics in American cities* (pp. 33-48). New York: Longman.

13

San Francisco: Postmaterialist Populism in a Global City

RICHARD E. DeLEON

SAN FRANCISCO'S FORMAL governmental structures scarcely have changed over the last half century; over the last two decades, nearly everything else has been a blur of transformation. Global and national forces have altered radically the city's population and have restructured its economic base. Unrestricted highrise development has "Manhattanized" the city's skyline and has expanded its central business district both up and out. The rapid competitive growth of nearby cities has diminished San Francisco's stature and clout in the region's economy. The federal government has withdrawn its visible hand of support and its concern for urban problems, while at the same time "tax reform" through Proposition 13 has whittled away the city's own revenue-generating options. These and other trends have degraded severely the city's quality of life: Traffic congestion has increased, with total gridlock on the rise; affordable housing is almost nil, and homelessness abounds; AIDS ravages the gay and minority communities, with crack cocaine reaping its share of victims as well. This is not the city Joseph Alioto knew when he took charge as mayor in 1967. It is the city Mayor Art Agnos must govern, if he can, into the 1990s.

San Francisco's problems as a city are not unique. What is unique is the city's high level of citizen activism and involvement in shaping local economic development and land-use policy and in pioneering social legislation in areas as diverse as domestic-partners legislation, comparable-worth programs, and cultural-arts planning. Perhaps more than in any other large American city, San Francisco residents have learned how to use the political system to control their own economic destiny and quality of life. This assertion of local popular sovereignty over widening domains of city life has come to be called *urban populism* (e.g., see Swanstrom, 1985). For reasons that will become clear later, I prefer the term *postmaterialist populism* in describing San Francisco's political transformation over the last 30 years. Whatever it is called, San Francisco's political culture is distinctive and has become a source of inspiration to citizen activists in other large cities, such as Seattle. The critical issue is one of longevity: How long can San Francisco's populist regime survive in a demographically turbulent and politically hyperpluralistic environment? That is one of the major themes this chapter will explore.

THE DEMOGRAPHICS OF DIVERSITY

San Francisco currently has a population of about 724,000, compared to 740,000 who lived there in 1960. That ranks it twelfth among the nation's largest cities, sixth among the 13 cities compared in this volume. (These and other comparative city statistics cited later are taken from U.S. Bureau of the Census, 1967, 1988.) Although the city's population has grown little in recent decades, its share of the rapidly growing Bay Area region's population (now at 5.7 million) has declined from 29% in 1950 to about 13% today. After New York City, San Francisco ranks first among large cities in population density: 16,142 persons per square mile. Given its very low residential vacancy rate (1.1% in 1988) and its scarcity of developable land (403 acres through the year 2005), San Francisco's scant 46.4 miles of space are unlikely to accommodate many more people in the decades ahead (see LeGates, Barton, Randlett, & Scott, 1989).

San Francisco's ethnic diversity is unmatched among the nation's large cities. According to a recent survey of all 3,000 U.S. counties, San Francisco ranked as the most ethnically diverse county in 1980 and is located in the most ethnically diverse metropolitan region in the country (Allen & Turner, 1988). Because immigration from Asian and Latin American countries continued through the 1980s and into the 1990s, the city's ethnic diversity actually is expected to increase. Of the city's total population, 48% was nonwhite in 1980, up from 18% in 1960. This percentage is certainly well over the 50% threshold in 1990, thus adding San Francisco to the growing list of cities with "minority majorities." The Asian/Pacific Islander population, incredibly diverse within itself, is the city's largest nonwhite ethnic minority group (23% of the total using 1980 Census data), followed by blacks (13%) and Hispanics (12%). Since 1980 the Asian and Hispanic populations have continued to grow, while the black population has declined in both absolute and relative terms. These more recent shifts in the city's ethnic

composition have prompted growing feelings of interethnic rivalry, particularly among blacks who feel they were there first in the postwar wave of "new
immigrants" and are now being displaced by the even newer immigrants from
Southeast Asia and Central America.

In the late 1960s the emergence of a large gay and lesbian community
added to the city's ethnic, cultural, and life-style diversity. Ranging in size
from 10% to 20% of the city's adult population, the gay and lesbian community remains politically active and economically potent, despite the tragic
attrition caused by over 5,000 deaths from AIDS.

San Francisco's ethnic complexion and life-style diversity produce a
population mix that in no way resembles the traditional image of mainstream
America. Barone and Ujifusa estimate that in San Francisco's 5th Congressional District, representing about 77% of all San Franciscans, "white non-
Hispanic, non-gay, native-born Americans make up only about 25% of the
population" (Barone & Ujifusa, 1989, p. 94). Traditional family life-styles
in San Francisco are now mainly the province of the nonwhite majority. In
1980, 49% of the city's whites 16 years or older lived in "nonfamily"
households, compared with 14%, 22%, and 27%, respectively, of Asians,
Hispanics, and blacks. Whites are much older on average and have fewer
school-aged children than members of these other ethnic groups. Further, the
white homeownership rate in 1980 (39.7%) just barely exceeded that of
blacks (39.1%) and Hispanics (38.2%); all three group homeownership rates
were eclipsed by that of Asians (49.6%). (These and other comparative
statistics cited later for the city's ethnic groups are based upon an analysis of
the U.S. Census Bureau's 1980 Public-Use Microdata Sample for San
Francisco. See DeLeon, 1991.) These demographic facts account for some
of the conflicts between the nonwhite community, especially Asians and
blacks, and the predominately white nonfamily gay/lesbian community on
such issues as domestic partners legislation and AIDS prevention. Indeed, a
major controversy continues in San Francisco over the very definition of
what "family" means, and there remains much opposition to gay and lesbian
efforts to attain legal recognition of same-sex marriages and nontraditional
family arrangements.

Relative to other large cities, San Francisco's population is well educated,
affluent, and highly skilled: 28.2% of those 25 years or older in 1980 had
completed 16 years or more of school (ranking first among the 13 cities
compared in this volume); money income per capita in 1985 was $13,575
(also ranking first), and 19.2% of those employed in 1980 held professional
or technical positions (ranking third behind Seattle and Boston). These
indicators suggest that San Francisco's population has a high level of what
Inglehart calls "cognitive mobilization"—the verbal, analytical, and commu-

nication skills needed to cope with social and political complexity (Inglehart, 1990, pp. 337-340). These overall figures conceal substantial social-class disparities, particularly among ethnic groups. By most indicators of socio-economic status (SES), San Francisco's non-Hispanic whites as a group are at the top of the SES ladder, blacks and Hispanics at the bottom, with Asian/Pacific Islanders occupying the middle rungs. To illustrate the magnitude of some of these disparities, comparisons of non-Hispanic whites and blacks show that 59% of whites 16 years or older had completed at least some college, versus 36% of blacks; 15% of whites earned incomes below 125% of the official poverty line, versus 28% of blacks; 35% of white employees were in managerial or professional occupations, versus 15% of blacks; 42% of white income earners derived income from interest, dividends, rent, or royalties versus 10% of blacks. The comparable figures for Hispanics are very similar to those for blacks, with Asians intermediate. As we shall see, these social-class differences and income polarizations are largely the consequence of San Francisco's changing economic base and occupational structure.

THE LOCAL HOURGLASS ECONOMY

Over the last two decades, San Francisco's economic base has become increasingly specialized in tourism and business services. This trend is bifurcating the city's occupational structure by expanding *both* the top tier of high-paying professional and managerial jobs *and* the bottom tier of low-paying personal services jobs. The supply of middle-income jobs is shrinking, thus explaining the "hourglass" image. To illustrate the magnitude of change that has occurred, in 1963 about 38,000 city jobs were in the services sector; by 1982 that number had mushroomed to over 100,000. Recent studies project that the number of low-income service jobs will increase by 43% between 1980 and 2005, while professional/technical jobs in finance, insurance, and real estate will increase by 18%; these two categories alone account for 74% of the projected increase of 119,500 jobs in San Francisco over the 25-year period (ABAG, 1987). Alarmed by these projections, a widely respected urban-planning association recently warned of the "danger of becoming primarily a service-oriented economy. . . . Instead of creating jobs that sustain a growing middle class population, we could experience growth in high paying business service occupations, such as lawyers and accountants, and the kinds of low paying jobs protected by the state" (SPUR, 1987, p. 1).

The days have long since passed when San Francisco served as the economic hub of the Bay Area region. The city has not had a majority of the region's jobs since 1947; Santa Clara County alone generated half the region's job growth over the last decade. The continuing deconcentration of population and jobs throughout the region has shifted the economic spotlight to other areas—e.g., San Jose (which now matches San Francisco in population, is building its own downtown, and has much more room to grow), Oakland (with its superior port facilities), and the entire Tri-Valley area just over the East Bay hills. In many ways, San Francisco has been reduced to the region's "symbolic" center and a convenient source of cultural amenities, fine restaurants, and venture capital to fuel the region's real economic engines in San Jose and Silicon Valley.

Many city leaders lament the passing of an age and long to restore San Francisco to a regional stature of at least first among equals. They particularly resent the city's gradual transformation into an urban amusement park. Approximately 1 in 11 city jobs is generated by nonprofit or for-profit arts organizations; 1 in 9 city jobs is in the tourism and "hospitality" industry. Fisherman's Wharf now caters to tourists, with hardly a fisherman in sight. These trends concern some analysts who see economic danger in such dependence upon the kindness (and money) of strangers. The city's declining port industries may yet revive through the efforts of the new port director to upgrade port facilities, to develop commercially 7 miles of dilapidated waterfront, and to attract new shipping business. Similar proposals to revive the city's manufacturing industry sound hollow, however, since the manufacturing sector has never been a major source of jobs and is not well suited to the limited office space and small floorplates permitted under the city's new growth-control regulations.

Arising from these trends and patterns are some of the key policy issues that have dominated San Francisco politics over the last two decades: What economic role will the city play in the rapidly growing metropolitan area? To what extent is highrise office construction necessary to attract the kinds of businesses and jobs needed to diversify the local economy and to restore the city's middle class? How may the crisis in housing affordability be resolved before it prices even the middle class out of San Francisco? What should be the balance between neighborhood preservation and commercial development? Should promotion of many small businesses be preferred to the relentless competition with other cities for a few large corporations? Should the city intensify economic development in areas where it already excels (e.g., business services, arts and design, biogenetic research) or invest major new resources to resuscitate laggard sectors (e.g., manufacturing)? These economic development and land-use issues are at the root of major conflicts between the city's traditional business elites and the new urban populists.

THE "CAGE OF AUTHORITY" AND THE "MOBILE OF GOVERNANCE"

Even after 16 years, Frederick Wirt's (1974) metaphors have currency as descriptions of San Francisco's highly detailed restrictive charter and its fragmented, loosely coupled government structures. In 1932, voters approved a special charter establishing a rare form of consolidated city and county government. (City and county territorial boundaries are identical.) A product of the Progressive Era, the 1932 charter was conceived as an instrument to safeguard local government against power grabs by corrupt politicians, especially those inclined to meddle in routine administration and licensing decisions. The charter put cages around all points of power and severely restricted discretionary authority. The original charter was itself a corpulent document, but it has now grown obese from the countless voter-approved amendments required to change such minutiae as job classifications or salary schemes. Despite frequent complaints about the "entropification" of the political process caused by the glut of proposed charter amendments that voters must wade through each election—25 such amendments in the 1988 election alone—voters have chosen repeatedly to live with this system by rejecting every major proposal for charter reform.

Through recombinant political splicing, framers of the 1932 charter created a truly novel constitutional monster. They melded elements of strong-mayor, city-manager, and commission systems into a strange government hybrid that must be unique among American cities.

Executive authority is divided between an independently elected mayor and a chief administrative officer (CAO) who, although appointed by the mayor, may be removed only by two-thirds vote of the Board of Supervisors or by recall. The mayor has appointive, budgetary, and veto powers, but administrative control is restricted to just those agencies and departments outside the CAO's domain (e.g., police and fire). Even within this limited sphere the mayor's leadership is filtered through a layer of commissions and boards (e.g., Police Commission, Planning Commission, Board of Permit Appeals), whose members are appointed by the mayor and in only half the cases serve at his or her pleasure. Because the mayor lacks formal executive authority over their powers, these boards and commissions exercise considerable autonomy in making policy and decisions. Mayoral appointments to such other agencies and commissions as the Redevelopment Agency and the

Port Commission must be confirmed by the Supervisors. The controller, like the CAO, is appointed by the mayor and operates independently of the CAO, thus dividing fiscal management functions. The offices of assessor and city attorney, typically staffed by appointment in other cities, are elective positions in San Francisco and highly prized by ambitious local politicians.

Overall, executive authority in San Francisco's City Hall is divided, dispersed, and decentralized. Wirt's image of a mobile with "figures frozen in midair and interconnected in inexplicable ways by wildly zooming lines" (Wirt, 1974, p. 11) is fairly descriptive of the city's governmental organization chart. It is a structure that resists centralized authority or efforts to employ it as an instrument of social change. The human resources are there—over 34,000 local government employees in 1982—but charter-ordained cages keep them separated and incapable of achieving a political critical mass. To a significant degree, Wirt's earlier conclusion that what San Francisco has is "government by clerks" still applies, as does his observation that the position of mayor "is what he makes it by the force of his character and personality" (Wirt, 1974, p. 13). In his campaign for mayor, Art Agnos spelled out an ambitious agenda in his booklet "Getting Things Done." But "getting things done" will require much more than his formal powers as mayor, which are modest at best, and will hinge instead on his skills as a political entrepreneur in building coalitions, assembling resources, negotiating deals, and harnessing the energies of government clerks.

Legislative authority under the 1932 charter was placed in a Board of Supervisors whose 11 members are elected at-large to staggered four-year terms. The Board initiates legislation, shares authority with the mayor over the budget (currently at $2.3 billion, up from $238 million in 1964-1965), may place proposed charter amendments on the ballot, must confirm some mayoral appointees and has a veto on removal of others, and provides a forum for public debate during regular sessions and committee hearings. One important thing the Board cannot do under the charter is to interfere in the administrative functions of the executive branches or to exercise effective oversight in the implementation of ordinances.

Despite the legislative responsibilities implied by a multibillion dollar budget and a complex government apparatus, the voters refuse to raise supervisors' salaries above the current $23,924 a year, making them the worst-paid county supervisors in the nine-county Bay Area. Evidently, San Franciscans continue to harbor the myth that running the city is a part-time job for public-spirited amateurs. The myth has consequences. Mayor Agnos soon will be losing one member of his liberal Board majority, Nancy Walker, who is resigning simply because she cannot afford the financial sacrifice any longer. Other members of the Board either must treat their work as a part-time job, indebt themselves to financial contributors, or be independently wealthy. Nevertheless, every two years, from 20 to 30 candidates run for the Board against incumbents who almost always run for reelection and rarely are dislodged. Frustrated by the sticky grip of incumbency on reelection campaigns and the free-for-all politics that dominate Board proceedings, voters in 1990 passed a citizen-initiated ballot proposition restricting incumbents to a maximum of two terms.

During the period 1977 to 1980, the electorate flipped back and forth through several elections in deciding between district versus at-large representation (see Hartman, 1984). In 1980, following a brief but consequential two-year experiment with district elections, the voters restored at-large elections and have stuck with them since. One reason for resisting a return to district elections is that the Board in recent years has become demographically if not substantively representative of the population under at-large elections: five women, six men; a gay person; two blacks, one Latino, and one Asian. Moreover, for the first time in memory, the Board has at least a 6 to 5 liberal/progressive majority that can facilitate enactment of Mayor Agnos's programs. The attitude of "if it ain't broke, don't fix it" may prevail. In view of the recent passage of the two-term limit, however, an initiative for district elections might succeed once again. It is still an item on Mayor Agnos's agenda, and the populist movement in San Francisco is still gathering steam.

The 1978 to 1980 interlude with district elections was consequential because, among other things, it attracted a crowd of new candidates who lacked the financial backing or name recognition usually required to win in an at-large election. One of them was Harvey Milk, who won in his district as the city's first openly gay elected official; another was Dan White, who won in his district on a wave of antigay sentiment and conservative resentment toward the new liberal mayor, George Moscone. The eventual tragic outcome of this conjuncture is now well known. On November 27, 1978, Dan White shot and killed both Harvey Milk and George Moscone. Following a controversial trial, White was found guilty only of voluntary manslaughter and given a light prison sentence. This incited the "White Night" riot of protest and rage, which in turn further deepened a cleavage between white working-class homeowners and the gay/lesbian community that persists to this day. (Dan White committed suicide shortly after being released from prison several years later.) This whole sad chapter in San Francisco's political history probably also delayed fulfillment of Mayor Moscone's progressive agenda through the nine-year moderate progrowth "interregnum" of Mayor Dianne Feinstein's regime until Art Agnos won election as mayor in 1987.

As in most Western cities, San Francisco's voters are endowed with the tools of direct democracy: the initiative, referendum, and recall. In recent years, these tools have been used by citizens both to advance and retard the sweep of populist reforms. Three examples will illustrate. Starting in 1971, the city's slow-growth activists used a series of growth-control initiative campaigns to circumvent progrowth resistance in City Hall by taking their case directly to the people for a vote. Abetted by growth-induced deteriorations in the quality of life, these campaigns cumulatively eroded the legitimacy of progrowth business ideologies, prompted significant preemptive growth-control measures from City Hall, and eventually culminated in the passage of Proposition M in 1986—the most sweeping and restrictive growth-control measure ever enacted by a large U.S. city (see DeLeon & Powell, 1989). On the other hand, the city's conservative groups also know how to employ the tools of direct democracy. In 1989 the Board of Supervisors unanimously passed and the mayor signed a "Domestic Partners" ordinance allowing unmarried couples who live together to register their relationship at City Hall. City workers who signed up under the law would receive many of the same rights, such as hospital visitation leaves, accorded to married couples. A week before the new ordinance was to take effect, a petition-gathering campaign organized by religious leaders and conservative groups succeeded in stopping implementation of the new law until it was put to the voters as a referendum on the November 1989 ballot. This citizen-initiated referendum measure, Proposition S, was contested fiercely and eventually lost by a slim margin. The pioneering ordinance was rendered null and void. As a final example of direct democracy at its worst, a petition drive by maverick leftists in 1983 did succeed in subjecting then-Mayor Feinstein to a special recall election. But the maneuver backfired: Feinstein received 86% voter support in a landslide victory that discouraged strong candidates from opposing her in the general election, thus solidifying her regime for another four years (see Hartman, 1984).

Local elections in San Francisco are officially nonpartisan. Yet the Democratic Party creates the partisan air that San Franciscans breathe, and in state and national elections the city belongs to the Democrats. Republican registration of voters is at 18.4%, down from 20.6% in 1981, and registered Democrats outnumber Republicans more than 3 to 1. The Democratic Party's local organizations (e.g., the county committee) are weak and exert little direct influence on city politics. This may change, however, as a result of recent court decisions allowing political-party endorsements of candidates in local races. Further, the local political-power vacuum created by the deaths of Representative Philip Burton in 1983 and of his wife and successor, Sala Burton, in 1987, is quickly filling up again with old and new Democrats working in concert to coordinate policies and to anoint candidates in local races. Most notable in this group are Mayor Art Agnos; Representative Nancy Pelosi, who succeeded the Burtons in representing the 5th District; Assembly Speaker Willie Brown; and Assemblyman John Burton, Philip Burton's brother. This local assemblage of liberal Democratic power, combined with stunning populist victories in growth-control legislation, led one analyst to conclude recently that San Francisco is the "temporary capital of the liberal wing of the Democratic Party in the United States" (Starr, 1988, p. 44). Nancy Pelosi adds her claim that San Francisco "has become the capitol [sic] of the progressive political movement in this country" (Starr, 1988, p. 157).

HYPERPLURALISM REVISITED: CHALLENGE AND OPPORTUNITY

An attempt to identify "the" power structure in San Francisco politics is as futile today as it was in 1971, when Frederick Wirt aptly described the city's highly granulated interest group politics as "hyperpluralistic" (Wirt, 1971). Two trends have intensified the city's hyperpluralism: the disintegration of the business-dominated progrowth coalition, and the political mobilization of ethnic minorities. These trends have leveled the political playing field, historically tilted in favor of downtown business elites and whites.

During the 1960s and 1970s, downtown business elites and their labor union allies formed a progrowth coalition to guide the city's economic development and land-use policies (see Mollenkopf, 1983; Hartman, 1984). In more recent years, however, that relatively solid center has been ground into smaller pieces by economic trends, slow-growth opposition, and the divisive impacts of Proposition M. The city's business community has lost its capacity to speak with a single political voice and to mobilize resources in leveraging a grid-locked world. Commercial real estate investors and landlords find that their interests diverge from those of developers and architects. Small businesses are not meshing well politically with large corporations. There is much confusion and lack of leadership within the business community in deciding what strategic options to pursue in coping with the new slow-growth policy environment. San Francisco's labor organizations have not rushed in to play a central leadership role in city politics. One reason is that labor's own core constituency is disintegrating. Unionization of the Bay Area workforce has fallen from a high of 38% in 1960 to a low of 20% in 1987. Only 6% of the burgeoning mass of service workers has been unionized. Like the business elites, labor leaders are turning inward to assess the damage and to recalibrate political strategies to fit the new environment and economic trends.

Just as economic changes and slow-growth opposition have undermined the progrowth coalition of business and labor, the demographic trends discussed earlier have eroded the hegemonic rule of white politicians. Fueled by immigration and rapid ethnic diversification over the last 30 years, growing numbers of blacks, Latinos, and Asians have mobilized electorally to achieve greater political clout and representation at City Hall. In 1962, for example, only 4% of the city's commission appointments were black or Hispanic minorities; by 1976 this figure had jumped to 30%, and it has risen even more under the Agnos administration (see Browning, Marshall, & Tabb, 1984). From mere token representation of minorities on the Board of Supervisors in the 1960s, the late 1980s found two blacks, one Latino, and one Asian all standing for election to the Board and winning on their own terms. Successful mobilization and group empowerment have made San Francisco's political institutions more closely reflect the ethnic and cultural diversity of the city's population. This is progress in the struggle for racial justice and equality, but the price paid has been an increase in hyperpluralism, further complicating the task of coalition formation and political leadership.

In sum, hyperpluralism is alive and well in San Francisco. If anything, the disintegrative forces of political disorder are even greater now than they were 30 years ago. The challenge to governance is formidable, and it remains to be seen whether Mayor Agnos's leadership and progressive vision can chart a course for the city amidst all the social turbulence and political disarray. Yet with the challenge come opportunities. Beneath the surface of San Francisco politics new alignments, coalitions, and ideologies are forming slowly, signaling the deeper restructuring associated with regime transformation. Citizens, politicians, and business leaders are redefining their roles and are relating to each other in new ways in making public policy. The generating source of this restructuring is the populist slow-growth movement and its postmaterialist value perspectives on the meaning and use of urban space.

POSTMATERIALIST POPULISM AND THE FORMATION OF A COMMERCIAL REPUBLIC

Postmaterialism is a term used by Ronald Inglehart (1990) to describe a new political polarization found increasingly in advanced industrial societies. It is a value-based political cleavage that cuts across traditional class divisions. It is focused on issues relating more to the meaning and quality of life than to its materialistic basis in property and productivity. These kinds of issues are of greatest concern to the younger generation, particularly those from middle-class backgrounds who are accustomed to affluence yet are critical of the harmful environmental impacts and marginal diminishing utility of economic growth. In Inglehart's view, global culture shifts are slowly recalibrating the traditional "Left" versus "Right" vocabulary so as to coincide with and give familiar meaning to this new axis of political polarization (see Inglehart, 1990). In San Francisco, perhaps more than in any other large American city, the polarities of the "materialist" Left and "postmaterialist" Left have been politically aligned by the coalition-building efforts of the slow-growth movement.

San Francisco's slow-growth movement arose in the early 1970s from a grassroots network of neighborhood associations, environmentalist groups, and political clubs. During its formative stage, the slow-growth coalition's narrow social base consisted mainly of white middle-class professionals who were politically motivated by quality-of-life concerns and neighborhood preservation goals. In mounting their first growth-control initiative campaigns, slow-growth leaders paid little attention to the material interests of low-income workers, renters, and people of color. Their stance was purely negative vis-à-vis the highrise development projects promoted by downtown business elites. These early campaigns failed at the ballot box.

Learning from these defeats, slow-growth activists expanded their electoral coalition base to include blacks, working-class homeowners, and struggling renters. They pressured City Hall into placing development in the service of redistribution through linkage fees, housing funds, job-training programs, and developer concessions. In later, more sophisticated initiative campaigns they articulated an alternative positive vision of the city's commercial public interest. They also sharpened their attacks on progrowth ideology, this time emphasizing not only the growth-induced deterioration in the quality of life, but also the failure of unrestricted growth to deliver on promises of decent jobs and affordable housing. This strategic balancing of materialist and postmaterialist goals eventually proved successful in building a winning coalition. In 1986, against weakening business resistance, the slow-growth campaign won majority-vo⸱ ⸱r support for Proposition M, which imposed permanent citywide caps on highrise construction and established priority policies restricting commercial land-use development. It became the new law of the "land" regulating private business use of urban public space. The slow-growth coalition consolidated its victory in 1987 by helping elect Art Agnos as mayor, thus establishing a City Hall connection and ensuring effective implementation of the new regulations. Over the last four years, Proposition M has become the written constitution of urban populism in San Francisco and an inspiration to slow-growth leaders in other cities and states.

What is important about these slow-growth campaigns is that they cumulatively *politicized* economic development and land-use issues. In most American cities, the key decisions regarding these issues are made by business leaders seeking profits and responding to market forces. Yet these decisions are central to the city's physical survival, economic vitality, and communal identity. By institutionalizing the exercise of popular sovereignty over these kinds of issues and decisions, San Francisco has moved beyond mere reform to achieve regime transformation. A new *kind* of city is being hatched in San Francisco—a "commercial republic" in which people and places have power and business power is put in its place to serve public goals (see Elkin, 1987).

Slow-growth movements now abound, especially in the Western states. What is distinctive about San Francisco's slow-growth movement is that it deliberately has embraced ethnic minority and working-class interests, pursued a political strategy of institutionalizing reforms and legitimating new social goals, and democratized the process of local economic development and land-use decision making. This is *populism* in its grassroots style, communal rhetoric, and stress on people before profits. It is a *middle-class* populism because of the high level of cognitive mobilization and technical skill required to give the people voice in managing their own local economy. It is also an *urban* populism, but as a social movement it may transcend local and urban boundaries to become a force in national politics. Adapting Inglehart's terminology, I prefer the term *postmaterialist* populism to describe what I believe is a global transformation of values viewed in local context, a condensation and congealing of "culture shift" forces in a particular place at a particular time. San Francisco is the place, and the 1990s could be the time when materialist and postmaterialist ideologies knot together to form a new kind of citizen, one truly capable of thinking globally and acting locally to solve urban problems. The political coalition that supports this synthesis of values is brittle and unstable. But if it can have time to jell within the institutional framework of Proposition M, and if Mayor Agnos can succeed in governing and "getting things done" through a difficult transition, San Francisco might become the nation's first full-fledged commercial republic practicing local economic democracy.

REFERENCES

Allen, J. P., & Turner, E. (1988, April). *The most ethnically diverse places in the United States.* Paper presented at the meeting of the Association of American Geographers, Phoenix, AZ.

Association of Bay Area Governments (ABAG). (1987). *Projections—87: Forecasts for the San Francisco Bay Area to the year 2005.* San Francisco: Author.

Barone, M., & Ujifusa, G. (1989). *Almanac of American politics 1990*. Washington, DC: National Journal.

Browning, R. P., Marshall, D. R., & Tabb, D. H. (1984). *Protest is not enough: The struggle of blacks and Hispanics for equality in urban politics*. Berkeley: University of California Press.

DeLeon, R. E. (1991). The progressive urban regime: Ethnic coalitions in San Francisco. In B. Jackson & M. Preston (Eds.), *Racial and ethnic politics in California*. Berkeley: Institute for Governmental Studies.

DeLeon, R. E., & Powell, S. S. (1989). Growth control and electoral politics in San Francisco: The triumph of urban populism. *Western Political Quarterly, 42*, 307-331.

Elkin, S. (1987). *City and regime in the American republic*. Chicago: University of Chicago Press.

Hartman, C. (1984). *The transformation of San Francisco*. Totowa, NJ: Rowman and Allanheld.

Inglehart, R. (1990). *Culture shift in advanced industrial society*. Princeton, NJ: Princeton University Press.

Legates, R., Barton, S., Randlett, V., & Scott, S. (1989). *BAYFAX: The 1989 San Francisco Bay Area land use and housing data book* (Report No. 89-10). San Francisco: San Francisco State University, Public Research Institute and the Bay Area Council.

Mollenkopf, J. (1983). *The contested city*. Princeton, NJ: Princeton University Press.

San Francisco Planning and Urban Research Association (SPUR). (1987). *Vitality or stagnation? Shaping San Francisco's economic destiny* (Report No. 234, p. 1). San Francisco: Author.

Starr, K. (1988, January/February). Art Agnos and the paradoxes of power. *San Francisco Magazine*, pp. 40-44, 157.

Swanstrom, T. (1985). *The crisis of growth politics: Cleveland, Kucinich, and the promise of urban populism*. Philadelphia: Temple University Press.

U.S. Bureau of the Census. (1967). *County and city data book, 1967*. Washington, DC: Government Printing Office.

U.S. Bureau of the Census. (1988). *County and city data book, 1988*. Washington, DC: Government Printing Office.

Wirt, F. (1971, April). Alioto and the politics of hyperpluralism. *Transaction 7*, pp. 46-55.

Wirt, F. (1974). *Power in the city: Decision making in San Francisco*. Berkeley: University of California Press.

14

Seattle: Grassroots Politics Shaping the Environment

MARGARET T. GORDON
HUBERT G. LOCKE
LAURIE McCUTCHEON
WILLIAM B. STAFFORD

IN 1965 EDWARD BANFIELD wrote that Seattle politics, far from being radical, corrupt, or bitter, were "downright dull." He described Seattleites as "busy making money, rearing children, trimming lawns and boating." Banfield saw prosperity and a high rate of single-family home ownership "giving Seattle a suburban quality." The populace, which then included 8.4% minorities, usually elected Republicans with business backgrounds to the weak-mayor, weak-council form of city government. Banfield portrayed the decision-making process at the time as a luncheon of the "Big Ten" business/civic leaders who reigned through citizen committees, causing him to ask whether there was really *anyone* in charge (Banfield, 1965, pp. 133-146).

Since by 1965 Seattle was well on its way to becoming a more cosmopolitan city, many argue that Banfield's portrait were already dated when it was published. Furthermore, city politics were changing at the time, from what Clarence Stone (1989) would describe as a rather tightly held "corporate" political regime to a more open and participatory political process. Since 1965, there have been major changes in the structure of government, the people, and the decision-making process.

AUTHORS' NOTE: Authors are listed in alphabetical order and made equal contributions to the chapter. We are grateful to Brewster Denny, James Ellis, Dan Evans, Phyllis Lamphere, Ken Lowthian, Richard Page, and Chris Smith for their comments on earlier drafts.

- Today, 25% of Seattle's population is nonwhite, which means a greater diversity of interests is involved in city politics. The greatest population growth has been among Asian Americans.
- Seattle school enrollments peaked in 1962 with over 100,000 students, but have declined to 42,500 in 1990 with about 55% being minority children. Although legally separate from the city, public education has become a major focus of city politics.
- Although city elections are by law nonpartisan, all mayors elected since 1969 have been Democrats, and currently all state legislators from Seattle are Democrats.
- Citizen activism brought about changes in the city charter and state law enabling a strong-mayor, strong-council form of government.
- Seattle's real and perceived power in the region has declined dramatically as population growth has increased in areas outside the city, giving rise to tensions between the city, suburban, and county governments.
- Citizen activism—both neighborhood and special interest—has come to dominate politics.

Perhaps most important, the self-image of Seattle has changed. The pre-World's Fair (1962) Seattle was viewed by many Seattleites and much of the rest of the country as a parochial, isolated, resource-bound city that was part of the last frontier—a "cultural dustbin," according to Sir Thomas Beecham, noted symphony conductor of the time (P. Lamphere, personal communication, August, 1990). The physical and cultural effects of the public capital investments by the city in the 1960s became evident in the 1970s and 1980s, and today Seattle views itself, as do others increasingly, as a cosmopolitan city capable of hosting such international events as world trade conferences and the Goodwill Games. Poised on the Pacific Rim, forging agreements with Canada, and close to Europe through polar routes, Seattleites struggle with new prosperity and growth as the city increasingly finds itself at the center of three emerging giant trading blocks of the new world economy—Europe, Asia, and North America. Furthermore, in 1990 Seattleites believe they count. Because they think they can affect outcomes, citizens actively participate in formal and informal politics, and the processes of identifying and solving problems seem as important as the solutions. Seattle politics today seem far from dull.

Proof for some observers that the current political dynamic is working well for the city exists in the results of several recent polls and studies rating Seattle as the nation's "best place to live" ("Best Place," 1989), the "most liveable city" (*Places Rated Almanac,* 1989), one of two or three "best places to raise children" (Ephron et al., 1988), the second "best place to visit" (*Condé Nast Traveler,* 1990), the "best city for women" (Ephron et al., 1988), and the second "best place to start a new business" ("How Seattle stacks up," 1990). Jokes about the famous 1970s billboard asking "Would the last person leaving Seattle please turn out the lights?" have been transformed into facetious stories and suggestions about how to discourage Easterners and Californians (especially) from migrating to Seattle while nonetheless claiming the mantle, City of Goodwill.

Seattleites are proud their city pulled out of the economic and related psychological depressions of the early 1970s and the early 1980s. They are now determined to be part of a process that establishes sustainable, environmentally sound, economic growth in the area. They remain intensely conscious of the beauty of both the city and the region, and they have learned that the "Rainier factor," as they call it, and the still comparatively low housing prices (though increasing rapidly in very recent years) mean immigrants to the region often are willing to accept reductions in salaries in order to live in the area. Seattleites are proud that their city, county, and state are often recognized for innovative solutions to problems, such as the city's recycling program.[1]

On election night, hosted by the local Municipal League, opposing candidates for a wide range of state and local offices, along with representatives of the press, gather together in a large hotel ballroom in a party atmosphere to monitor voting results. The scene is unimaginable in Chicago or Boston or Philadelphia. And Seattleites are proud that their city, with only 10% African Americans, in the fall of 1989 elected a mayor of African-American descent, Norm Rice, by a near landslide of 57% over a well-qualified white opponent.

All this is not to say that Seattle has no problems. It does, and the two that seem most pressing are: (a) the nature and quality of education throughout the city's school system, and (b) the host of issues related to rapid growth in the area (traffic congestion, skyrocketing housing prices, fear of overbuilding lest there be another economic slump, shortage of solid-waste disposal sites, increased crime). Concern exists that an entrenched underclass may be developing, with related problems of homelessness and drug abuse. A wide range of people, however, individually and in grassroots organizations, is working to solve these problems.

The transformation in Seattle from what Banfield saw as the dull, dreary days of the 1950s and 1960s to what many see as the vibrant, cautiously hopeful 1990s is the result of a fortunate confluence of local and national events and economic trends, together with creative and cooperative civic and political leaders willing to work energetically for the common good. In this chapter we describe the major changes in Seattle and Seattle politics since the early 1960s. These include shifts in the demographic characteristics of the city, changes in structure of city and county government, shifts in party alliances and political regimes, waxing and waning of the role of business leaders in city politics, changes in the aggressiveness and role of the press, and growth and role of citizen interest groups and neighborhood organizations.

THE LOCATION, THE ECONOMICS, AND THE DEMOGRAPHICS

Although blessed with great natural beauty, Seattle is situated in relative isolation from the rest of the country, and so it traditionally has been outside the mainstreams of migration and commerce that have shaped the characters of cities in the East, Midwest, and South. This geographical isolation is related also to a sense of political and cultural isolation and a feeling that, since they are a long way from the commercial and political centers of the country, Seattleites have to look out for themselves. From its earliest days, Seattle has looked to Alaska and Asia as important trading centers. Since World War II, there has been a developing awareness worldwide of the increasing economic and political strength of Asia and a subtle shifting of the frontiers westward, making Seattle more central and less isolated than ever before.

One of the unique features of Seattle is that, more than most other cities, Seattle's postwar economy has been tied to a single corporation—in this case the Boeing Company, one of the world's largest manufacturers. Unlike other corporate towns traditionally dominated by a single company, Seattle has felt Boeing's political influence only indirectly. It has been focused instead at the state level, especially with respect to taxation issues, environmental regulations, and economic development. Seattle learned in the 1970s that when Boeing, as part of a cyclic industry, slumped or boomed, the economic effects would be felt acutely in the city.

During the past 50 years, the fortunes of the airplane giant have plummeted twice, in 1969-1971 and in 1982-1984. In the first "Boeing Bust," the number of Boeing employees in the state fell from 102,400 (a fifth of Seattle's population at the time) to 41,000 three years later. Since the multiplier effect of a Boeing job then was about 2.8 (for every 1 Boeing job, another 1.8 jobs were created indirectly), the effects of the bust were very widespread. Corporate taxes plummeted too, which led to severe budget cuts in most state-supported programs. Many workers sought jobs elsewhere, giving rise to the then-popular joke about turning out the lights. Boeing, and with it Seattle, again suffered during the recession of the early 1980s, when the number of employees fell from 79,500 in 1980 to 59,800 in 1983. Since that time, Boeing has made a spectacular comeback with a backlog of airplane orders that promises profits late into the 1990s. The company's employment reached an all-time high in 1989 of about 116,000 (Pascall, Pedersen, & Conway, 1989).

Despite Boeing's recent successes, the early 1980s recession gave local area residents the jitters, and the current optimism can be characterized only as cautious. There is talk about diversification and the need to establish sustainable economic development. Although the percentage of the region's labor force that worked for Boeing in the 1960s was greater than it is now (25% as opposed to 18%), the multiplier effect of a Boeing job rose from 2.8 in the 1960s to 3.8 by 1989. Thus, although the region's direct dependency upon Boeing for jobs has diminished, the stronger multiplier means that Boeing remains as important in the region's economy as it ever was. (Pascall et al., 1989.)

While the region's other major businesses, employers, and institutions of higher learning also suffered their own turbulence in the 1980s, by the end of the decade most had experienced influxes of new funds and employees, which allowed them to share in the cautious optimism. Adding to the education-dependent enterprises in the greater Seattle area has been the rise of the software giant, Microsoft, in suburban Redmond. Already exceeding a billion dollars in sales annually, Microsoft recently announced it will double its workforce to about 8,000 in the near future ("Microsoft Planning," 1990).

Despite the concentration of growth in areas outside the city proper, Seattle's downtown business environment has remained strong. Indeed, the number of jobs in Seattle (both within and outside the CBD) doubled from 268,000 in 1960 to approximately 537,000 in 1988. Virtually all the city's major office buildings have been built since the mid-1970s. Between 1983 and 1990, local, national, and foreign entrepreneurs added a total of 9.34 million gross square feet of office space to the downtown. Despite the attractiveness of some of the new buildings, the rate of construction was so great in the latter 1980s that citizens passed an initiative with 70% of the vote to "cap" the number of square feet of office space that could be built each year in the central business district.

The growth has generated a regional groundswell of public concern over escalating housing prices, traffic congestion, loss of wetlands and open spaces, and general environmental deterioration. The effective, environmentally sound management of sustainable growth has thus become a major focus of politics in the area. As Seattle's dominance of the region in terms of population wanes and, therefore, its political clout diminishes relative to the rest of the surrounding area, it is likely that Seattle's government will direct

TABLE 14.1
Seattle and King County Population, 1960-1990

	1960	*1970*	*1980*	*1990*
King County	935,014	1,156,633	1,269,749	1,499,509*
Seattle	557,087	530,860	493,846	516,000**
% Seattle	60	46	39	34

SOURCES: Decennial Censuses for Washington State; U.S. Bureau of the Census.
*Unofficial, partial, and preliminary counts from the 1990 post-Census local review program (September, 1990).
**1990 Statistical Abstract of the United States, Table 40, pp. 34-36. Washington, DC: U.S. Government Printing Office.

increasing efforts toward strengthening cooperative interrelationships with other local governments and the establishment of some kind of regional governing body.

The economic ups and downs in Seattle have been accompanied by changes in the Seattle population. The population in the city proper declined from 557,000 in 1960 to a low of 494,000 in 1980 and rose again to an estimated 516,000 in 1990. (See Table 14.1.) Meanwhile, the remainder of King County surrounding Seattle grew by 190,000 people to 965,000 during the same time period, with most of the growth in the last decade. Thus, in 1990 only about a third of the county's population resides in Seattle. The lack of growth in the population of Seattle, however, conceals much more dramatic changes in the characteristics of its inhabitants.

One of the unusual demographic features of Seattle—perhaps due, in part, to its geographic isolation—is the ethnic and racial character of its populace. A city with strong Scandinavian heritage, it also has African American and Asian communities whose roots are traceable to the beginning of the century. Unlike other West Coast cities, however, Seattle's African American community received its last great influx immediately following World War II; four decades later, the black populace remains a relatively small percentage of city residents. The Native American population in Seattle likewise has not changed in any significant way since the period of the Banfield analysis.

In 1960 the city was 92% white, with a substantial proportion being of Scandinavian descent. The remaining 8% were split about equally between African-Americans and Asians, with a few Native Americans. By 1980, minority representation had grown to 20% and included 3% of Hispanic origin. In the last decade, Seattle has been a major receiving port for Asian immigrants, and the 1990 Census is expected to show a substantial increase in that population.

Increases in the Asian (other than Japanese) and Hispanic populations, together with the presence of a large Filipino community, add to the diversity. In total, the nonwhite residents of Seattle in 1990 are estimated to constitute about one quarter of the city's population and slightly more than one half of the enrollment in the public schools (U.S. Bureau of the Census, personel communication, 1990).

In 1960, Seattle was very much a family-oriented city, with 26% of its residents below age 15, the result of the baby-boom years experienced with the rest of the nation. During the last three decades, however, largely because of declining fertility levels but also because of strong out-migration of families, Seattle has experienced a decline in the number of families with children. In 1980, only 14% of Seattle residents were under age 15, a trend that experts believe the 1990 Census will show continued during the 1980s. More so than most other cities in the country, Seattle was a net exporter of children, with many, especially whites, moving to the surrounding suburbs. Busing has been cited as a cause, although this pattern existed well before the city's voluntary desegregation busing program began in 1978, and the trend has continued. The out-migration of primarily white children and the in-migration of primarily Asian students has resulted in a dramatic increase in the proportion of minority children in the city schools, a phenomenon true of most if not all urban areas of the nation during the same 30 years.

In contrast to the out-migration of families, Seattle has been a net importer of young adults ages 20 to 29, particularly between 1970 and 1980. Local experts believe the 1990 Census will show the same patterns for the last decade. If correct, the city may be characterized appropriately as having a bi-modal age distribution; that is, it will be home for young adults and senior citizens, many of whom will live alone.

Undoubtedly, the most significant impact of the out-migration and current age distribution of the city has been on the Seattle Public School System. In 1962 enrollments peaked with over 100,000 students, and in 1990 school officials report a fall count of 42,500 (Seattle Public School District, personal communication, October 1990). In 1960, 92% of the students were white; in 1989 45% were white, with 24% African American, 21% Asian, 5% Hispanic, and 3% Native American. Approximately 45% of the students are now living with one parent, a factor associated with greater poverty.

The decline in enrollments has been accompanied by a perception of increasing deterioration in the quality of the K-12 education system. In 1990 the quality of public school education was identified by many as the most critical community issue. Up until the 1960s the Seattle School Board has been a "blue-ribbon" committee with political clout in the city. As subsequent school boards became more populist in nature, they seemed to have less political power and support in the city. A teachers' strike was followed by a record drop in enrollments of white students. In 1978 the Seattle public schools implemented a voluntary desegregation plan[2] and enrollments dropped again, causing crises over pending school closures. Between 1979 and 1985, 29 schools closed, 3 of which were to be relocated (Lamphere, 1989). Several citizens' groups turned their attention to desegregation, enrollments and related school closures, and the quality of education. In turn, these issues affected neighborhood politics and, eventually, city politics.

These issues became the major campaign issue for Mayor Norm Rice and his opponent, and in the months since Rice took office, a central focus of his energy has been to motivate citizens to participate in local "education summits" to search for solutions. This emphasis on education by the mayor is unusual in Seattle because legally the mayor has little or no control over the education system. The school board is elected separately, and education is funded by the state and voter-approved levies.

POLITICS AND CIVIC ACTION

As the economic and demographic forces have played themselves out during the last 30 years, Seattle, King County, and the Central Puget Sound Region also have witnessed significant changes in politics and governance. These three decades saw city politics transformed from a weak-mayor, weak-council, strong old-boys'-network to a strong-mayor, strong-council, watchful citizens' network. What provoked the transformation?

If Seattle politics have changed significantly—some would insist radically—since the era of Banfield's observations, it is in part a reflection of the city's predilection for progressivism rather than partisanship in its politics, an attitude and atmosphere that has marked the city's political life at least since World War II. Some attribute this to the large number of military personnel who passed through Seattle *en route* to the Asian theater and returned to settle after the war. Seattle's current "can do" style of politics and municipal problem solving is a part of the city's progressive political tradition. Others argue that the price of increased political vitality has become a pronounced tendency toward single-issue politics, NIMBY ("not in my back yard") postures, and an obsession with process in the city's public affairs.

The progressive spirit in American politics has been as much a way of thinking as it has been an era of political and social reform. Given Seattle's geographic isolation, progressivism came later to the City of Goodwill than to the rest of urban America and, perhaps for the same reason, has lasted longer. Progressivism was thus a dominant feature of Seattle city politics in the 1950s and 1960s, when city government was firmly in the hands of the progressive-minded business community that launched a host of capital improvements and started the environmental movement. It has come to stamp indelibly the political process in a city where matters of public policy quickly and clearly override party loyalties. If the absence of strong partisanship makes for dull politics, Seattle remains as politically uninteresting as it was when Banfield wrote about it in the mid-1960s. But if citizen activism over municipal and regional issues is the stuff of the urban political processes, if urban politics may be defined as the way things get done and not just what goes on in City Hall, then the Seattle style of municipal politics is generating lots of excitement.

Since 1960 Seattle has had six mayors: Gordon Clinton (1956-1964), Dorm Braman (1965-1968), Floyd Miller (appointed to fill Braman's unexpired term in 1969), Wes Uhlman (1970-1977), Charles Royer (1978-1989), and Norm Rice (1990-present). The first two of these mayors were tied closely to and worked with a small group of businesspeople who strongly influenced what happened in the city, the third served during the transition to the strong-mayor system, and the last three have had to engage an ever-widening and diversified range of citizens' groups in order to succeed with many important initiatives.

One of the major changes in the style of politics since 1965 is reflected in the role of the Municipal League. Established in 1910, the League traditionally engaged in campaigns urging better planning and a workable metropolitan government, and it vigorously took positions on most major issues facing the city (Crowley, 1985). The influence of the Municipal League declined, however, when the growth of neighborhood and issue-based citizens' groups shifted the focus from capital improvements to equity and justice issues. Toward the end of Royer's regime, women and minority activists became especially important. Whatever else was going on throughout the 30-year period, citizen involvement has included a steadfast focus on the degree to which economic growth is or is not compatible with the quality of life and the environment. Toward the end of the 1980s, the Municipal League began to reemerge as an important citizen-based force in the city and region. Led by its first woman president, B. J. Narver, the League engaged in studies and issued influential reports on the key issues facing the area.

THE TRANSITION TO THE 1990s

Some believe that the story of the transition from the Seattle Banfield was describing to the present day began when some young professionals decided to do something about the water in Lake Washington. It had turned brown and had begun to smell as pollution generated by the population growth after World War II threatened the large 28-mile-long lake separating Seattle from its Eastside suburbs. The lake warned citizens of King County that their governance system was not working very well. As with most voter-approved government reorganizations, it took events that citizens could touch or feel, or in this case, see and smell, rather than studies or blue-ribbon commission reports to prompt government restructuring that generated corrective action.

The brown water was caused by a number of lakeside municipalities and sewer districts that disposed of their raw sewage into this treasure. A citizens' group appointed in the mid-1950s by then-Mayor Gordon Clinton and King County's commissioners was given the challenge of developing a solution. It was chaired by a young attorney, James Ellis, who had earned his stripes as chair of a citizens' group that had attempted to reform the county charter in order to allow the county to better deal with cross-jurisdictional problems. Although the reform was rejected by voters in 1952, by 1957 awareness had grown about the need for coordinated countywide action, especially with regard to Lake Washington.

In 1957 Ellis and his colleagues convinced the Washington State Legislature to authorize the creation of METRO—The Municipality of Metropolitan Seattle. It was in theory to coordinate the region's sewer, transit, water, solid waste, parks, and planning functions. METRO's authorization won by one vote on the last night of the session. Patterned after the then recently created Toronto federated structure, METRO was to be composed of 16 members, including the mayor and members of the City Council of Seattle, elected King County officials, suburban and sewer district officials, and citizens. The following year, after two attempts, the voters activated the sewer functions (only) of METRO, and the cleanup of Lake Washington began. All of the $135-million tab was generated locally, and the plan became a national model for pollution abatement programs.

During the creation of METRO, Seattle and King County politics had been dominated by Republicans and conservative Democrats. The mayoralty was, by design, institutionally weak; the budget was prepared by and controlled by the City Council. The city operated through a multiplicity of boards and commissions and exuded a philosophy that government should do as little as possible. Ross Cunningham, then political editor of the *Seattle Times,* was quoted by Banfield as saying, "If you want to get anything done in Seattle, you get about six members of the Big Ten [business men, including Ellis] together and tell them it's a good project. If you can convince them, you're in" (Banfield, 1985, p. 141).

One of the last achievements of the Big Ten corporate regime was the vision and implementation of the 1962 World's Fair. Business leaders thought it would be good for business and would put Seattle on the map. Seattle seemed to awaken then, and the region has never been the same. The heart of the fairgrounds, now Seattle Center, is still the major center for arts and cultural events for the area, and the towering "needle" is still the symbolic logo for the city—like the arch of St. Louis, the Trans America Building of San Francisco, and Empire State Building of New York.

The positive attitude about Seattle's future following the Fair led Ellis to engage in yet another successful business-citizen-initiated effort. In 1965 Ellis delivered a speech entitled "Rapid Transit and the Shape of the City" (J. Ellis, personal communication, August, 1990). He challenged the community to take charge of its own accelerating growth through a comprehensive program of capital improvements, including rapid transit and new parks, roads and community centers. The proposal soon was augmented with ideas for a county domed stadium and other amenities. The whole package thrived under the banner known as Forward Thrust. While other cities were plagued by race riots and burning neighborhoods, Ellis and a Committee of One Hundred pushed Seattle voters to approve $334 million in bonds to pay for this massive package of capital projects in 1967.

The stories of the formation and activation of METRO in 1957, the inception and execution of the World's Fair in 1962, and the packaging and approval of Forward Thrust in 1967 capture the changes in citizen activism that fueled the political transition in Seattle from the 1960s to the 1990s.

Rather than the "corporate" (Stone, 1989) Big Ten regime, METRO was the work of a group of civic-minded Young Turks—Seattleites who returned from the war to find their city in the throes of physical deterioration, symbolized by the use of Lake Washington as the community cesspool. But cleaning up the lake didn't rattle any cages, so to speak; the Big Ten didn't mind if the lake got cleaned up (J. Ellis, personal communication, August 1990).

The World's Fair a few years later was again the effort of the Seattle Big Ten establishment. It was initiated by the businesspeople who were concerned about what they saw as the city's backwater image, and they decided unilaterally to do something about it.

By the time Forward Thrust came along in 1965, the political climate had changed. Ellis and others realized its success required a broader coalition within the city and the region. A Committee of One Hundred was created and they turned to the electorate to win passage of the bonds necessary to complete the underwriting of the host of capital improvements. Forward Thrust was a harbinger of the type of citizen activism that would come to characterize the new style of Seattle politics.

In addition to working on Forward Thrust, citizens in the mid-1960s got involved in changing the structure of government. Some activist, reformer citizens' groups in Seattle began to grow restive as civil rights marches and demonstrations elsewhere captured the attention and imagination of Seattle-ites who thought things should change in Seattle too. In the mid-1960s citizens helped Mayor Braman bring about the racial integration of the police and fire departments (B. Denny, personal communication, October 1990). In 1967 a citizens' group led by soon-to-be-elected city councilwoman Phyllis Lamphere pushed the state to change laws which, coupled with subsequent city charter changes,[3] enabled Seattle to move in 1968 to a strong-mayor, strong-council form of government. It was as if the electorate had decided that city government was too important to be left to the corporate powers-that-be. In 1969 the community elected a young, energetic Democrat, Wes Uhlman, as mayor. Uhlman had been the youngest legislator in the state's history and had held important chairmanships in state government. Elected in his early 30s, he defeated a much older, prominent Republican, Mort Frayn. Uhlman entered City Hall with a young political staff and with the agenda of consolidating power in a strong-mayor form of government. He also had an agenda for the arts and culture, services for the elderly, historic preservation, and a new human service role for the city through creative use of federal grants. The economic decline in the area at the same time forced Uhlman to expand the role of city government in order to deal with unem-ployment and related problems.

It was then that the City Council began to change dramatically. Fresh from success in changing the state's blue laws, three young lawyers formed a new organization, Choose an Effective City Council (CHECC). At the time, the average age of City Council members was 68. In 1967 CHECC selected a young woman Democrat, Lamphere, to run against an old-line Democrat, and a young Republican, Tim Hill, to run against an old-line Republican on a coalition ticket. The successful careers of the first woman president of the National League of Cities and the current county executive and former two-term City Council member thus were launched.

In the elections that followed, the City Council was transformed into a younger, stronger, more accountable body. Since being on the nine-member City Council is a full-time job, well paid by national standards, reasonably well staffed by professionals, and elected at-large, sometimes it is facetiously called the 10-mayor system.

In retrospect, the early 1970s are touted as the "golden age" of the Seattle City Council. The new progressive mayor and the new progressive City Council sought and got resources from the state, then led by Governor Dan Evans, and from the federal government, brought home then and later by Senators Warren Magnuson (chair of the U.S. Senate Appropriations Com-mittee) and Henry M. Jackson (chair of the U.S. Interior Committee). A decade of major change followed. The confluence of the economic crisis, brought on by the first Boeing Bust, and the availability of federal and state funds gave the new energetic political and civic coalition the opportunity to take on new responsibilities and to demonstrate its acumen.

In 1971 the City Council and an architect pushed historic preservation and launched the Save the (Pike Place) Market movement; in 1973 the Council began levying a 1% charge on all construction costs for city arts projects; it created one of the nation's first departments of Human Resources; and it enacted a water-quality act that became a national model.

As alluded to earlier, from the 1950s on activists and elected politicians had been urging some form of federated regional government to confront problems that crossed jurisdictions. To replace the weak three-county commissioner form of government, in 1969 a charter review committee recommended that county government should be reorganized. Ultimately, a county executive and a full-time county council were proposed, and the voters agreed. The nine-member County Council was elected by district. One of the three commissioners, Republican John Spellman, became the first county executive (and later was elected governor). Since the population of the unincorporated area then was approaching Seattle's, these County Council members felt pressure to provide urban-level services. The Council members representing Seattle and other incorporated areas, however, were more interested in the county providing regional services.

With the growth of county population outside of Seattle, the new county government became a far more aggressive player on the metropolitan field, leading in the 1970s to the first of many intergovernmental battles. The first dispute took place in the mid-1970s when King County led the withdrawal from the four-county Puget Sound Governmental Conference. Pierce and Snohomish counties followed suit. They argued that the organization was ineffective and that planning and decision making should take place at the county level, preferably by the county governments. Under the pressure of existing federal regulations a compromise was reached creating a two-tiered structure—a four-county Puget Sound Council of Governments (PSCOG), with four strengthened county subregional councils.

The County Council then turned its sights on METRO, which had succeeded in 1972 in adding public transit to its sewage functions during the fall election. By this time METRO had consolidated Seattle's transit department with that of a privately owned suburban bus company. METRO had improved the service and its own visibility and was at the height of its organizational popularity. The County Council was able to persuade the state legislature to enact a bill allowing, with voter approval, the county government's absorption of all METRO's functions. The county's argument was that its representative structure was more appropriate for METRO functions than the federated structure. Both political parties supported the county in its takeover attempt, but the proposal was defeated in 1978 by over 75% of the voters. The aggressiveness of the County Council on the PSCOG and METRO issues was to haunt city-county relations for years. Growth of the suburban cities led to their increasing clout on METRO. In 1989, helped by aggressive, hostile press coverage and a flap about the use of South African granite in building an underground bus tunnel; a radio purchase; and the long struggle over siting of a secondary treatment plant, county officials and supporters led a charge on METRO that resulted in the resignation of the executive director and the resurfacing of the governance question. During the summer of 1990, a U.S. District Court judge ruled that the system of representation for METRO was unconstitutional and ordered it to be fixed. Alternative proposals abound, and one longtime participant-observer argues that the only way to solve all the problems is to call a "freeholders' election" and to hold a Puget Sound constitutional convention, keeping the decision making close to the voters (Denny, 1990).

While the county struggled with METRO, the Uhlman administration consolidated the power of mayor by centralizing decision making under his position. This included budget making. During budget hearings, Uhlman ousted the fire chief for insubordination, igniting a recall campaign by the Seattle firefighters against Uhlman, who, having survived, chose not to run for reelection in 1977. That fall, Charles Royer defeated four pillars of the Seattle City Council, including Lamphere, and in January 1978 he began a 12-year reign as mayor. Royer was a young neighborhood-based Democrat who had been a successful television news commentator. He seemed to personify the completion of the transition from the Big Ten dominance of city politics to the young, progressive, strong-mayor, strong-council form that has dominated Seattle politics ever since.

Royer continued many of Uhlman's initiatives, and by 1980 Seattle had become a liberal, Democratic city. The progressive Republicans, including Dan Evans, who had added much to state and local politics, were gone. Randy Revelle, a Democrat, was elected in 1980 as county executive in a close race. Despite the similarity of Royer's and Revelle's views on many issues, city-county relations remained contentious. By the mid-1980s not 1 of the over 20 state legislators from Seattle was a Republican.

Metropolitan affairs were simply not a high priority for Royer in the early years of his administration. His focus was on issues and less on process at that time as he worried about the location for a new state convention center, whether and where to build a transit tunnel, replacement of the West Seattle bridge, saving the public health hospital, and planning downtown renovations that included Westlake Mall, a public square and shopping area. In defending his priorities in 1983, Mayor Royer said, "We want a downtown that is constantly alive and filled with people. Downtown should be the region's common ground, the neighborhood everyone owns" (Royer, 1983). From Royer's perspective, these issues were related to his main political concerns—health care for the poor, transportation, affordable housing, and a healthy downtown, including the waterfront and Seattle Center. Royer garnered national attention for himself and for Seattle when he advocated national urban causes as president of the National League of Cities when it was struggling with President Ronald Reagan's budget cuts and devolution of many federal programs to the states.

During the mid-1980s, as the region recovered from the second Boeing recession and the economy began to boom again, the most important issues became growth and gridlock, especially on the outskirts of Seattle. Vacant lots and greenbelts were becoming multifamily developments, while two-lane country roads were becoming *de facto* parking lots. Politicians were challenged at the polls on growth and quality-of-life issues. Suburban mayorships and City Councils, once dominated by the local Chambers of Commerce, gave way to the community activists who had gained political experience pushing an array of citizens' initiatives. The politics of scarcity shifted to the politics of plenty. Environmentalism took hold, and the city's voluntary residential recycling program achieved nearly 80% compliance and became a national model. Major political battles began to be fought over what residents *didn't* want in their towns or neighborhoods, be it a jail, a dump, or a new apartment house.

In 1989 Royer chose not to run for a fourth term and accepted a position as head of the Institute for Politics at the Kennedy School of Government at Harvard University. He didn't sell his home in Seattle and said on his departure that he hoped to return after "teaching people in the East how we do things in Seattle" (C. Royer, personal communication, May 17, 1989). The 1989 race was in many ways similar to that of 1977 and involved a number of contenders, including Doug Jewett, city attorney and a Republican. Jewett chose the schools and the busing program, mandatory for all residents, as the central issue of his campaign. He spearheaded the filing of the Save Our Schools initiative that would have eliminated busing and given the saved dollars to the schools. Jewett needed an issue that would split the Democratic and independent voters, and busing was it.

Jewett's surprise opponent was Norm Rice, an African-American City Council member who filed on the last possible day. Rice had lost to Royer in the 1985 mayoral race and to Jim McDermott for U.S. Representative in 1988. Observers thought Rice had decided to give up politics. He promised active city leadership on the school issue, including an education summit. The irony of the campaign was that, although the Save Our Schools initiative backed by Jewett passed narrowly, Jewett himself lost, and Rice was elected the first African-American mayor in Seattle's history. Rice thus became the third progressive Democratic mayor since 1970. Rice made it clear soon after his election that, in addition to public education, improving regional relationships was one of his highest priorities.

On matters concerning greater Seattle the state legislature passed a landmark growth-planning act that will change the nature of local, regional, and state decision making. Many observers believe local jurisdictions will not relinquish control unless the state requires regional planning. A state Growth Commission presented recommendations on governance and growth management to the governor for the 1991 legislative session (Ford, 1990). A new regional body is being proposed to implement the new legislation following a vote to "kill" the PSCOG by mid-1991. If adopted, the Growth Commission recommendations will offer a significant test of the capacity of local officials in a voluntary federation to make tough decisions. A citizens' initiative (#547) scheduled for the November ballot that would have repealed the Growth Commission, making its recommendations moot, and set up a state-controlled regional structure was defeated.

Predictably in Seattle, all this governmental activity is accompanied by substantial neighborhood and civic activity. The reinvigorated Municipal League has conducted studies and written position papers on the growth-related issues and the restructuring of government. The Greater Seattle Chamber of Commerce has planned leadership retreats. The local newspapers have identified both growth and schools as major issues on their editorial agendas, and the *Seattle Times* has hired a team of analysts to do regular features on growth in the region. PSCOG has also published its regional growth plan, Vision 2020.

The League of Women Voters of King County, in a 1978 study of governance, began their report with the statement, "Proposals for governmental reorganization in King County reappear with the regularity of the tides. However, the present surge of proposals for change indicates that there is a growing perception that issues exist which cannot be resolved under the present governmental structure" (League of Women Voters, 1978). After a decade of no action on the issue, the League updated the report in 1988 and changed "tides" to "seasons" (League of Women Voters, 1988). The 1988 report is likely to indicate substantial change has taken place or perhaps that "seasons" have become "glacial movements."

Our account of the changes in Seattle city politics leads to the conclusion that there has been a clear shift from a corporate political regime to a progressive alliance (Stone, 1989). Early on, the Seattle citizenry was content to let the Big Ten run the city because the Big Ten produced in the process some very commendable civic efforts that were good not only for business but also for the public weal. A younger, more diverse set of leaders took the reins in the 1960s and began to move toward a governance process in which the electorate had a much stronger voice. Some observers agree that while more and more citizen activism has developed over the years, questions remain about how constructive it is. Some wonder aloud whether of late that activism hasn't been too negative—saying No, stopping things—rather than positive, constructive building.

Whatever the answer to that question, during the last 15 years citizens have taken a very active role in changing the city charter; electing a mayor and City Council with different passions and commitments; shaping and reshaping government institutions; and more generally, experiencing the results of their power.

THE 1990s AND BEYOND

As Seattle moves toward the close of the century, it is faced with significant challenges. One of those is how the city and other local governments will choose to work together to solve cross-jurisdictional, regional problems facing them all. At the moment a vigorous struggle is taking place, with many concerned groups advancing their proposed plans and interests. The evidence from previous efforts—from METRO to Forward Thrust, and PSCOG—have produced relationships and processes that comprise a basis for successful, cooperative action. It remains to be seen whether Seattleites can create a workable model for interjurisdictional governance that other cities will want to emulate.

Seattle is perched upon the edge of the Pacific Rim, both a gateway to and a participant in the North Pacific and South Pacific trading groups. The state of Washington, largely because of Boeing and the Port of Seattle, has more trade with European nations than any other U.S. state. It may well be thrust into a pivotal position at the crossroads of European and Asian trade. Washington and Alaska both border Canada, and Washington has developed new agreements with its neighbors to enhance regional commerce and environmental protection, while Seattle and Vancouver have agreed to engage in some cooperative planning. Seattle's position of isolation in the last 150 years may in the future become a position of centrality as the Pacific Northwest becomes the Pacific Northeast of the next century.

The great physical beauty, the vigor and variety of arts and culture, the availability of spectator sports of all types, and the range of outdoor activities available to residents generates vitality and commitment to the area and causes a willingness to work to preserve the environment for the common good.

The politics are honest and open, scandals are rare, and elected officials seek citizen input. The people expect to be consulted before important decisions are made, and the politicians listen when civic organizations offer advice. The process is important to everyone involved; in fact, some have described modern Seattle as the "process capital of the world," sometimes causing "paralysis by analysis." The population is relatively well educated, and voters make informed choices. The most recent mayors, Royer and Rice, communicate well with the public. Some observers call them the "feel good" mayors. They both have an informal style, giving rise to feelings that city politics are as accessible as the beauty and the arts. Seattle, at the moment, appears to be practicing grassroots democracy as the framers of the U.S. Constitution intended. It is not dull at all.

NOTES

1. The City Solid Waste Management Department was selected a 1990 winner of the Ford Foundation/Harvard University Project on Innovations in State and Local Government and received a $100,000 award.

2. The city voluntarily developed and implemented a desegregation plan, but for individual students participation was mandatory. The implementation of the plan led to a statewide antibusing initiative being overturned by the courts.

3. These included "departmentalizing" city functions under the mayor (B. Denny, personal communication, October 1990).

REFERENCES

Banfield, E. C. (1965). *Big city politics.* New York: Random House.

Best place in America to live. (1989, September). *Money,* pp. 124-141.

Condé Nast Traveler. (1990, November), p. 204.

Crowley, W. (1985). *The Municipal League of Seattle and King County: 1910-1985.* Seattle: The Municipal League.

Denny, B. (1990, October 7). Metro fix is chance to fix whole county system. *Seattle Post-Intelligencer,* p. D3.

Ephron, D., Paretsky, S., Haskell, M., Kahn, A., Brown, R., Birnbach, L., & Laskas, M. (1988, September). "The best (and worst) cities for women." *Savvy,* p. 45.

Ford, R. (1990, July). *A growth strategy for Washington state: Preliminary discussion draft,* Washington State Strategies Commission.

How Seattle stacks up. 1990, February). *Sound Business,* p. 8.

Lamphere, P., Speech given at the Seattle Chamber of Commerce Leadership Conference, Semiahmoo, June 12, 1989.

League of Women Voters, King County (WA). (1978, February 19). *Is it time for a change? A study of governments in King County,* p. 1.

League of Women Voters, King County (WA). (1988, September). *Regional government study,* p.1.

Microsoft planning bigger workforce. (1990, June 16). *New York Times*, p. 29.

The Municipal League Foundation. *Issue Watch,* January-June, 1990.

Pascall, G., Pedersen, D. H., & Conway, R., Jr. (1989, September). *The Boeing Company Economic Impact Study.* Seattle: The Boeing Company.

Places Rated Almanac. (1989). Englewood Cliffs, NJ: Prentice Hall.

Stone, C. (1989). *Regime politics: Governing Atlanta 1946-1988.* Lawrence, KS: University Press of Kansas.

Royer, C. (1983, June). *State of the city address to City Council.*

U.S. Bureau of the Census. (1960, 1970, 1980). General population characteristics for Washington State. *Decennial census.* Washington, DC: Goverment Printing Office.

Washington Policy Notes. Institute for Governmental Research, University of Washington, 1978-1984.

15

Conclusion:
End of the Millennium Big City Politics

H. V. SAVITCH
JOHN CLAYTON THOMAS

LESSONS CAN BE GLEANED from our experience with 13 big cities, lessons about the continuing paradoxes that bedevil big cities. Our cities have at once become more important as centers of the nation's postindustrial might, yet more dispensable because of suburban alternatives. Big city mayors have become more important as public leaders, yet they are less able to lead divided cities. Thirty years ago the words *urban crisis* led the nation's political agenda. As urban problems have become more severe, they seem to have disappeared from the public mind.

How can we understand these paradoxes? Our explanation hinges on a series of developments, evidenced in the 13 profiles, that have affected and continue to shape these cities. These include (a) the growth of the exogenous city, (b) the rebuilding of local economies, (c) the metamorphosis of the urban crisis, (d) the continuing significance of race and ethnicity, and (e) the quest for central authority amid hyperpluralism.

THE GROWTH OF THE EXOGENOUS CITY

The political gravity of cities has shifted from mostly endogenous, or self-generating, sources to predominantly external, or exogenous, pulls. Cities are largely the product today of influences external to their boundaries. From the global economy to federal and state governments to the suburbs just beyond city boundaries, these exogenous influences say more about what cities are like and what they will become than do any factors within the boundaries.

The global economy has impinged upon local economies in ways Banfield could not foresee. The departure of factories, warehouses, and wharves did not leave all big cities empty. Instead, many big cities attempted to adapt to a changing world economy by pursuing corporate headquarters, by building international airports, and in a phrase, by seeking to situate themselves as the nerve centers of a postindustrial world economy (Sternlieb & Hughes, 1975; Savitch, 1988). Big cities are no longer just nodes for a regional or even a national economy but have grown into a much larger role. And, when the headquarters of major companies settle in the downtown, they bring with them new patterns of employment, new populations, new construction, and even a new culture.

Federal and state governments always had some impact on cities, but until the 1960s big cities were largely self-governing entities in most respects. Federal and state actions were few in number, limited in scope, functionally specific, and rarely accompanied by elaborate strings (Connery & Leach, 1960; Gelfand, 1975).

With the launching of Lyndon Johnson's war on poverty, federal urban policy took a quantum leap into a new intimate involvement with urban problems (Lowi, 1969). State policies often followed in the federal wake; as the federal involvement became embracing, so too did the state involvement. Washington and the states were no longer just helping pay for welfare or highway costs but were reviewing plans, requiring matching funds, insisting on citizen participation, monitoring spending, and, most importantly, determining local priorities.

The extent of this influence is suggested by the proportions of big city funding that came from federal and state sources. As Table 15.1 shows, by the mid-1970s federal aid to big cities had escalated dramatically. Without exception every one of the 13 cities hooked into the Washington pipeline, and every urban budget depended upon this largesse. The state picture, as shown in Table 15.2, is a bit spottier, but with few exceptions most big cities also took advantage of increasing state support. We should recognize, too, that in these decades urban budgets were expanding, making the absolute amounts in external aid even more impressive.

Once Reagan-era policies took full effect in the mid-1980s, the flow was less abundant. With cities now accustomed to this support, when federal aid no longer sufficed, the states were asked to make up the difference. For the most part, however, that plea went unanswered as state proportions of local funding also declined slightly, on average, during the 1980s.

TABLE 15.1
Federal Aid to Cities: 1960-1988 (as percentage of a city's general revenue)

City	1960	1968	1976	1984	1988
Atlanta	.79	2.34	11.13	12.84	7.04
Boston	.00	5.14	18.18	9.58	2.59
Chicago	3.01	13.47	13.65	21.05	10.94
Denver	.69	.67	14.24	7.00	4.01
Detroit	1.52	8.95	27.46	12.47	5.90
Houston	.63	3.50	15.78	7.52	10.54
Los Angeles	.68	.68	14.25	11.18	2.94
Miami	.00	.24	8.73	11.11	8.87
New Orleans	3.33	2.26	27.23	18.40	11.88
Philadelphia	4.16	11.28	23.16	9.96	4.06
St. Louis	.73	1.76	16.85	17.11	9.63
San Francisco	2.65	6.21	17.57	7.82	4.66
Seattle	.21	1.19	12.79	6.23	3.21

SOURCE:For 1960 figures: Bureau of the Census. (1960). *Compendium of city government finances in 1960* (G-CF60-No. 2), Table 5. For 1968, 1976, 1984, & 1988 figures: Washington, DC: Government Printing Office. U.S. Bureau of the Census.(Year). *City government finances in (Year)* (GF, No. 4), Table 5. Washington, DC: Government Printing Office.

TABLE 15.2
State Aid to Cities: 1960-1988 (as percentage of a city's general revenue)

City	1960	1968	1976	1984	1988
Atlanta	7.89	8.44	11.42	4.04	1.81
Boston	23.71	35.33	23.96	41.82	44.00
Chicago	12.75	8.36	11.65	12.83	15.62
Denver	32.08	28.25	18.48	13.11	17.87
Detroit	20.27	13.80	15.12	23.30	32.14
Houston	.00	.39	1.30	2.01	.82
Los Angeles	9.28	14.62	11.21	6.10	9.09
Miami	.00	.24	17.45	12.07	9.85
New Orleans	18.10	13.01	15.12	10.51	5.26
Philadelphia	6.39	7.16	14.81	12.98	16.96
St. Louis	2.93	6.64	11.57	8.03	6.07
San Francisco	22.65	25.56	25.72	25.57	25.34
Seattle	16.04	16.55	11.35	15.68	11.16

SOURCE: For 1960 figures: U.S. Bureau of the Census. (1960). *Compendium of city government finances in 1960* (G-CF60-No. 2), Table 5. Washington, DC: Government Printing Office. For 1968, 1976, 1984, & 1988 figures: U.S. Bureau of the Census. (Year). *City government finances in (Year)* (GF, No. 4), Table 5. Washington, DC: Government Printing Office.

Decreased funding did not necessarily mean reduced influence because federal and state influence increasingly found expression through new forms. Federal tax incentives spurred many urban development initiatives, such as gentrification in Boston and Philadelphia. Federal and state courts showed a new interest in local politics, with court orders directing the course of school desegregation in Boston and St. Louis and ending the tradition of patronage-hiring in Chicago. State politics also figured in new growth controls adopted in Seattle, San Francisco, and Los Angeles.

Suburbs, too, exert much more influence on big cities today than they did 30 years ago. Prior to the 1960s large companies unhesitatingly settled in downtown business districts. By the middle of the decade many companies were packing off to the suburbs. Center cities were forced to compete for industry with their once-forgotten hinterlands. It became commonplace for surrounding towns to raid big cities, not only for their businesses but for their most affluent residents, their sports teams, and their political prowess. As never before, cities became vulnerable to the suburban lure, always affected by what suburban growth could do to them and rarely able to respond effectively. From the city's point of view, it was more dependent upon forces outside its boundaries (Savitch, 1979; Kantor, 1988). On the other side of the issue, as long as suburbs could take the best from cities without paying a price, why should legislators for suburban constituencies care?

Even when businesses remained downtown, their employees often moved to the suburbs. Their commuting into and out of the city changed the city's social and political character. People who worked in the city but did not live in it often cared less about it. With the unity of work, residence, pleasure, and place shattered, urban citizenship was bound to suffer, and it did. Crime, unsuitable housing, and corruption became more tolerable because they could be left behind after working-hours.

THE REBUILDING OF LOCAL ECONOMIES

These exogenous forces figured prominently in the rebuilding of local economies that preoccupied big cities in the 1970s and 1980s. Cities sought to redevelop their central areas, in particular, both to accommodate the international order and to meet the suburban threat. The strategies they conceived to achieve this revival were shaped invariably by the incentives and funding available from federal and state governments.

The accelerating competition for development—among cities, as well as between city and suburb—generated a host of ingenious ways to attract capital. Urban-development corporations cleared land, arranged zoning requirements, and put together financial packages to facilitate new business. Cities went into partnership with private developers to build office buildings. They entered into complex long-lease and buy-back agreements with private syndicates.

The impact on big cities was geometric, sharpening distinctions in the urban physical form, contouring its politics, and deepening fissures in its social structure. Physically, the results were sometimes exhilarating, sometimes disappointing. The successes are familiar enough—skyscrapers in San Francisco and Chicago, convention halls in Atlanta and Detroit, sports stadia in Houston and Denver, waterfronts laced with shops and aquariums in Boston and New Orleans. But failure was also common: the fortress-like Renaissance Center in Detroit, reflecting that city's continuing lethargy; the many gleaming but empty office buildings in Houston and Denver.

All this was bound to change city politics. Mayors, city councils, and other local leaders were only too happy to welcome new investors. As a result, big city politics took a sharp turn away from a concern with local services and patronage, toward an emphasis on land-use zoning, property tax abatements, and bonds for economic development.

For a while the politics of land development encountered little citizen opposition. But as one development piled upon another, neighborhoods eventually responded. Depending upon the city, the response took different forms and brought different results.

In high-growth, prosperous cities, trenches were dug between downtown, established progrowth coalitions versus neighborhood, populist antigrowth coalitions (Mollenkopf, 1983; Logan & Molotch, 1987). In such cities as San Francisco and Seattle the conflict eventually brought voter approval of moratoria against office development or legislation of growth controls. So long as Boston was growing, it too found creative ways of addressing the problem by charging surtaxes for office development or by linking high-profit development to support moderate and low-cost housing.

In hard-pressed cities, where virtually no one opposed growth, neighborhoods fought instead for a share of the limited funding available to support development. Such struggling cities as Chicago and New Orleans swung back and forth between downtown developers and neighborhoods. Chicago featured the contrast of Mayors Daley (the senior), Bilandic, and Byrne, who courted downtown development, versus Mayors Washington and Daley (the junior), who tried to reconcile that development with neighborhood needs. The most devastated city, Detroit, had little choice but to auction its land to interested investors, as in the notorious case of Poletown, where a community of 3,000 residents was swept away to build an auto plant.

Neighborhoods in these cities also have faced direct threats to their indigenous fabric. As upper-income groups took jobs in downtown office towers, some moved into old neighborhoods, often with government encouragement and assistance. *Gentrification* and *displacement* became popular terms in the lexicon of urban planners.

A great debate ensued over whether gentrification was good or bad for the city. Proponents could point to Philadelphia's Queen Village, Chicago's Oldtown, or New Orleans's Garden District and see signs of renaissance. At the same time, housing prices in some cities skyrocketed, and longtime low-income residents often were displaced to unknown destinations.

In spite of a newly invigorated economy and its investment spillover, problems in many other neighborhoods nagged or even worsened. As the profiles in this volume have shown, some of the cities experiencing the greatest downtown resurgence continued to suffer neighborhood decline. Segregation increased, poverty spread, and neighborhoods became more dangerous. Clearly, downtown development and gentrification had their limits. Atlanta might boast 21 companies on the *Fortune* 500 list and consulates from 39 countries, but most of its residential areas are walled off from the central business district by highways and hostile streets. Much of San Francisco might float on a rising economic tide of Asian trade, but some city neighborhoods had run aground.

THE METAMORPHOSIS OF THE URBAN CRISIS

In the 1950s urban problems had just begun to reach public awareness. Poverty, racial segregation, family breakdown, and declining educational performance had started to crack through a shell of public indifference. By the late 1960s these problems had become critical, as evidenced dramatically in the urban riots of that period. Yet, by the 1970s and 1980s people either no longer cared, no longer believed they could do anything to help, or denied any problem existed. The Nixon White House exemplified this changing outlook. In his first term, Nixon declared that "we face an urban crisis" and established a special office for urban affairs. At the start of his second term in 1973, Nixon eliminated the office and announced that "the hour of crisis has passed" (*National Journal Reports,* 1973).

Ironically, the worst was yet to come. The problems of the 1950s only foretold new generations of problems in succeeding decades. Welfare dependency and chronic unemployment were added to the urban crisis in the 1960s, to be joined by crime and fiscal calamity in the 1970s. In the 1980s drug addiction, teenage pregnancy, the feminization of poverty, and AIDS also

TABLE 15.3
Employment in Big Cities: 1960-1986 (in thousands and percentages)

City	Employment		Change (%)	Unemployment(%)	
	1960	*1986*	*1960-86*	*1960*	*1986*
Atlanta	197	210	6.6	3.6	7.5
Boston	288	278	-3.5	5.0	4.4
Chicago	1502	1264	-15.9	5.4	9.3
Denver	196	261	33.2	3.6	7.2
Detroit	612	407	-33.5	9.9	11.8
Houston	364	862	136.8	4.3	10.9
Los Angeles	1014	1535	51.4	6.5	7.5
Miami	126	179	42.1	7.3	8.3
New Orleans	224	214	-4.5	5.6	11.2
Philadelphia	789	679	-13.9	6.5	6.9
St. Louis	294	179	-39.1	5.4	9.4
San Francisco	331	378	14.2	6.2	5.5
Seattle	230	280	21.7	6.1	6.8

SOURCE: U.S. Bureau of the Census. (1962). *County and city data book 1962*, Table 6. Washington, DC: Government Printing Office. U.S. Bureau of the Census. (1988). *County and city data book 1988*, Table C. Washington, DC: Government Printing Office.

weighed heavily upon big cities. By the 1990s cities were faced not with a single crisis but with layers of interrelated crises always seemingly about to explode. In Detroit that explosive potential has been institutionalized with a Halloween perversity called Devil's Night, during which arsonists stalk the city, setting hundreds of buildings ablaze.

At the same time, not all big cities have been hit equally hard. Table 15.3 suggests the differential impact on cities from the perspective of employment.

Over the period of a quarter century, older cities in the Midwest and Northeast suffered the most. Detroit and St. Louis saw the number of jobs plummet between 1960 and 1984. Chicago, Philadelphia, and Boston also lost jobs. The well publicized postindustrial office expansion in the latter two cities apparently was not enough to compensate for lost jobs in manufacture.

By contrast, newer cities in the South, the West, and on the Pacific Rim have done very well. Houston stands out as the premier growth center, with employment growth well in excess of what any other city experienced. (The lustre of this reputation has since been tarnished by Houston's recent economic downturn, as evidenced by the 1986 unemployment rate.) Employment also grew substantially in Denver, Los Angeles, and Miami.

Those data are consistent with other impressions. For the most part, newer big cities in the West and Pacific Rim (San Francisco, Los Angeles, Seattle,

TABLE 15.4

Crimes Against Persons in Big Cities: 1989 (per ten thousand of population)

	Murder +	Rape +	1989 Robbery +	Assault =	Total	1960-89 % Increase
Atlanta	.58	1.62	15.94	21.38	39.51	2079
Boston	.17	.83	10.11	11.16	22.27	1005
Chicago	.25	.00	10.57	12.59	23.41	279
Denver	.11	.66	2.56	3.89	7.22	135
Detroit	.60	1.37	11.45	10.59	24.01	323
Houston	.27	.67	5.73	4.73	11.40	326
Los Angeles	.25	.58	9.03	12.60	22.46	274
Miami	.35	.75	19.12	16.75	36.97	475
New Orleans	.47	.73	10.31	7.78	19.30	594
Philadelphia	.29	.47	6.19	3.97	10.93	225
St. Louis	.39	.81	10.42	19.59	31.21	407
San Francisco	.10	.51	6.66	6.13	13.40	232
Seattle	.07	.93	4.76	7.61	13.37	956

SOURCE: *Sourcebook of Criminal Justice Statistics*, 1990. U.S. Department of Justice, Office of Justice Programs, Bureau of Justice Statistics, Table 3.119, pp. 440-444.

Denver, Houston) have incurred less damage from the urban crisis than older cities in the East (Boston, Philadelphia) and especially the Midwest (Detroit, St. Louis, Chicago). The picture in at least one Southern city (New Orleans) is less encouraging.

While such economic issues as employment are important, social issues also defined city distress. This was particularly true of crime, which eroded the urban base and caused families and businesses to flee to safer suburbs. Not all big cities are the same, and some are safer than others. We might expect that prosperous cities would have less crime than those where employment opportunities are limited, but this is not always the case. More important, urban safety is a relative question because every big city has experienced soaring crime rates.

Table 15.4 shows how big cities compare with one another over the last quarter century. The table is limited to crimes against persons (murder, rape, robbery, assault) and does not include crimes against property (burglary, theft, arson). In order to draw some composite comparisons, the four categories of crimes against persons are added up and compared against a unit of population.

Note the economically harder-pressed big cities (Detroit, St. Louis) are also hardest hit on the incidence of crime. Surprisingly, these Midwestern cities are joined by two relatively well-off Southern counterparts (Miami, Atlanta). The "best off" cities include Denver, Philadelphia, Houston, San Francisco, and Seattle. Even with this category, it is astonishing to find increases of between 135% and 956%. Clearly, economic growth did nothing to stem the crime wave, and in some growth cities (Los Angeles, Atlanta) the wave has risen to the level of a great inundation.

THE CONTINUING SIGNIFICANCE OF RACE AND ETHNICITY

Underlying the continuing urban crisis is a pervasive and dangerous split between white and black America. A quarter of a century after the Kerner Commission warned that cities were "moving toward two societies, one black, one white—separate and unequal" (National Advisory Commission on Civil Disorders, 1968, p. 1), the movement continues and the wounds of antagonism have been rubbed raw.

As whites have moved to suburbs, a large black underclass has been left behind in big cities, increasingly separated by physical space. Income differences by race are just as compelling. Although the average income of blacks has increased, the gap between white and black households has widened. While poverty has not appreciably worsened for the population as a whole, three times as many blacks as whites live in poverty (Bernstein, 1988, p. B8).

The polarization between the races is also psychological. Just 10 years after the Kerner Commission Report, black perception of white society had become more negative. More blacks believed that they were denied employment because of discrimination, that racial prejudice had grown worse, and that whites were hostile to black advancement ("Racial Attitudes," 1968; New York Times/CBS News Poll, 1978). At the other end of the spectrum, more whites object to further action by government to change discriminatory practices, oppose scatter site housing, and attribute poverty among blacks to an underclass culture (Harris & Wilkins, 1988).

These attitudes help explain the numerous incidents of overt racial conflict reported in the cities examined here. The Miami riots of the 1980s were rooted deeply in a sense of racial injustice. The effort in Boston to establish a separate black enclave, Mandela City, hints that some blacks are ready to "secede" from white society. The racial slurs of a Houston City Council member illustrate the unfortunate persistence of racial stereotypes among some city leaders.

The picture is not all gloomy. During the last 30 years a substantial number of blacks have created their own middle class, often moving to suburbs.

Blacks have increased steadily their electoral representation in city councils, state legislatures, and in Congress.

Most notably, blacks have reached the top in many city halls. When Banfield wrote, none of these 13 cities had a black mayor. Currently, elected black mayors lead six of the cities. The list of black mayors, past and present, includes such well-known figures as Andrew Young, Coleman Young, Tom Bradley, Harold Washington, Sidney Barthelemy, Wilson Goode, and Norman Rice. Black mayors have even been elected in some cities, such as Seattle and Los Angeles, where whites constituted a majority of the voting population.

Distinct from the issue of race is the role of ethnicity in big cities. Ethnic solidarity is not new to big cities. Irish, Italian, Jewish, and Eastern European immigrants made the most of bloc voting and political patronage, and used ethnic and kinship ladders to enhance business opportunities. Many contemporary big cities have continued that tradition with new immigrant populations. The most formidable example is the Cuban-American population of Miami, which has combined political and business solidarity. In Denver, Federico Peña's Hispanic identity has helped in mobilizing voters, although he, like some black mayors, could be elected only by building a multiracial coalition.

The ascension of minorities to mayoral posts may hold more symbolic than substantive significance. Black mayors, for example, usually have changed the complexion of power, but they rarely have challenged its uses. Tom Bradley, Coleman Young, Wilson Goode, and most other black mayors have been well within the mainstream of big city politics. The same can be said of Hispanic mayors Maurice Ferre, Xavier Suarez, and Federico Peña. Still, the success of minority mayors in running multiracial cities could yet constitute a critical element in the improvement of race relations.

Ethnic consciousness and its political ramifications probably will become more visible in the 1990s and beyond. Ethnicity has been a durable commodity, and big cities continue to become more rather than less ethnic. St. Louis, for example, is on the verge of becoming another majority black city; Houston may be heading toward eventual "majority minority" status. There may be no greater challenge for tomorrow's big cities than that of building consensus despite growing ethnic diversity.

THE QUEST FOR CENTRAL AUTHORITY AMID HYPERPLURALISM

Issues of central authority figure as prominently in big city politics today as they did in Banfield's time, but the focus of the debate has shifted dramatically. Where Banfield and his contemporaries asked mostly whether authority was too *centralized,* observers of the 13 cities in this volume increasingly ask whether that authority has instead become too *decentralized.*

The questions about overcentralization were sparked by Floyd Hunter's (1953) study of Atlanta politics. Following in a long tradition of power structure studies (Lynd & Lynd, 1929, 1937; Hollingshead, 1949; Vidich & Bensman, 1958), Hunter argued that Atlanta epitomized rule by a "power elite," that is, a small number of business leaders who were able to translate economic power into political power. In alliance with political leaders (who might themselves have come from the business ranks), this elite dominated important decisions about the city. In the Atlanta case, a handful of corporate executives led by Coca Cola supposedly cooperated with Mayor William Hartsfield in ruling the city.

Similar power elites may be found in the profiles of other cities in this volume in the 1960s and before. As a few examples, Detroit had "The Big Three," the leaders of the three major automobile corporations; Miami had its Non-Group; Seattle had its "Big Ten" economic leaders; Houston had its "Suite 8F crowd"; Philadelphia had its gentlemanly elite and old "Main Line" business establishment that were the engine of the reform movement.

Not everyone saw big city politics as so centralized. Dahl (1961) argued by contrast that many cities were run by plural elites, groups that coalesced around particular issues. Rather than a single business elite deciding all issues, the groups that held power supposedly varied by issue. Other observers pointed to other elements of decentralization. Sayre and Kaufman (1961) described a multicentered political system in New York City. Truman pointed out that interest groups operated in a context of shifting equilibria (Truman, 1960).

These traits were evident in many cities in Banfield's (1965) volume, even some of those with apparent power elites. Seattle was said to be governed "by a network of little administrative hierarchies, each with a council committee at its apex" (Banfield, 1965, p. 137). Governing Detroit supposedly required "maintain[ing] a balance between bitterly antagonistic interests" (Banfield, 1965, p. 51).

No matter how fluid the power arrangements, the big city of the 1950s and early 1960s seldom lacked effective central authority. These cities invariably possessed a structure of power that anchored their coalitions and assured continuity between regimes. Homegrown industry, locally owned newspapers, and local political parties provided the basis for regime continuity.

That stable political infrastructure of an earlier era has withered or disappeared in most big cities. Homegrown industries have been replaced by multinational conglomerates or nonindigenous companies whose attachment to a city rests upon tax incentives and more sunshine. Locally owned newspapers are being bought by national chains. Even local parties that were rooted firmly in working-class wards have been sapped of their strength. The tradition of Richard J. Daley as chief party broker has given way to Richard M. Daley as political explorer who searches for a dependable constituency.

At the same time, other elements of the political composition of big cities have splintered—into a polarized white and black electorate, into neighborhood self-protection, and into highly contentious public-employee unions. Even business classes have splintered—sometimes by ethnic distinctions such as the Anglo-Cuban split in Miami, other times by geographical splits between suburban developers and downtown interests.

The splits brought to cities an intensification of the old pluralism. As distinguished from its forebear, the new hyperpluralism draws sustenance from the separation between the parts of the body politic; these pieces go their own way in seeking narrow gains; they put fewer restraints on their conduct; and, they are less apt to accommodate to one another.

This hyperpluralism bears a strong kinship to Yates's "street fighting pluralism" with its "unstructured, multilateral conflicts" and its "fragmented unstable and reactive policy outcomes" (Yates, 1977). In contrast to the Yates version of the 1960s and early 1970s, however, contemporary hyperpluralism has moved from the streets of the city to the corridors of city hall, transitioning from the pitched battles of "street-fighting pluralism" to periodic negotiations for a share of the pie in the "corridors of power." With that transition has come a muting of the stridency that characterized the earlier hyperpluralism but no lessening of the demands on the resources of the city.

As a consequence, few, if any, of the elite power structures of the Floyd Hunter era survive today. Alone among these 13 cities, Houston continues to be run by economic elites in much the same fashion as 30 years ago. Even there, economic diversification has expanded the city's economic leadership so much that it can no longer fit in the single "Suite 8F" hotel room.

The more typical city today may be Philadelphia, where the same fragmenting forces that gained power in the 1960s continue to pull apart the city in the 1990s. As one notable change, also perhaps typical of contemporary big cities, city-suburban hostility exerts an increased centrifugal pull over what it did in any time past. The results of the fragmentation are alarming; a growing city debt, culminating in the city's bonds recently being rated at the level of a junk bond.

On the positive side, these splintered forces have gotten together on occasion. Big cities have proved creative in assembling effective governing regimes despite the pulls of hyperpluralism. In the 1970s new coalitions formed in some cities between progressive political leaders, neighborhood and/or minority organizations, and the federal government, sometimes without the participation of local business. Business, traditionally the *sine qua non* of urban governing regimes, became expendable as a regime partner because the federal government was available to bankroll the progressive coalitions. In Los Angeles, for example, new Mayor Tom Bradley forged such a coalition between blacks, white liberals, the emerging environmental movement, national Democratic leaders, and federal funding agencies.

That alliance was no longer possible once Ronald Reagan arrived at the White House, but any reversion to the traditional business-dominated governing regime of the 1950s was equally unthinkable in the ethnically diverse city of the 1980s. Business was again an essential element to an effective governing regime, but it had to find new partners in order to have the necessary electoral support. In perhaps the best-known case of regime transformation, downtown business in Atlanta reestablished its authority by allying with the city's growing black middle class and Mayor Andrew Young. With Young's departure as mayor, it remains to be seen whether Maynard Jackson's return to the mayoralty will continue that coalition.

Most recently a few cities have developed a new form of environmentally conscious progressive regime. These coalitions join some business interests, often excluding businesses that favor unrestricted growth, with a quality-of-life oriented middle class in support of a slow-growth, proenvironment public agenda. To be sure, the pattern is restricted to growing cities—in particular, the Pacific Rim cities of San Francisco, Seattle, and perhaps Los Angeles.

With their roots in hyperpluralism, the post-1970 political coalitions have tended to be short-lived and *ad hoc*, in large part because they are so disparate and internally conflicted. As a consequence, they often lack the stability of the governing regimes of the 1950s and 1960s.

In understanding the rising currents of hyperpluralism it may be useful to put the idea in chronological and comparative context. In considering this context, two variables stand out as critical: the strength of political leadership and the cohesion of business elites. Examining cities in this way underscores the linkage between politics and economics, between public and private sectors, and provides us with a focal point on how a political regime operates at these vital intersections. Thus, the elitist model formulated by Hunter and others (Hollingshead, 1949; Lynd & Lynd, 1929; Mills, 1959) sees local political economies run by a hierarchical and cohesive private sector, whose top business executives set agendas for relatively weak political actors. At the opposite pole, the pluralist model developed by Dahl and others (Banfield, 1961; Sayre & Kaufman, 1961) stressed the strength of public leaders (Dahl's "executive-centered coalition") in bringing together a congeries of mixed, often diffuse, and sometimes competitive actors from the private sector. Now consider Yates hyperpluralist model, which sees political attenuation and economic diffuseness everywhere. Neither political leaders nor private actors are powerful enough to pull together the strings of the urban political economy. Add still another model developed by Schmitter (1977) and others (Savitch, 1988; Simmie, 1981) that posits both strong political leaders, who set their own agenda, and a unified business elite, who knows what it wants. Under this model, called corporatist, powerful actors in both public and private sectors guide the city toward specific ends.

Each of these models can be used to understand the political regimes of our 13 cities. The models are drawn from predominant trends within a given city and ought not to be interpreted in starkly black and white or absolute terms. Rather, elitist, pluralist, hyperpluralist, and corporatist systems can best be thought of as "ideal types," enabling us to think about the direction and evolution of cities. The idea here is not so much to furnish pat answers about which cities belong in what categories, but to focus on two variables and ask just how strong are political leaders vis-á-vis economic elites and why have cities spawned particular regimes.[1]

Figure 15.1 plots these models along the two axes of political leadership and economic elites. Examples from past regimes are indicated by dates in parentheses, while contemporary regimes have no dates indicated.

Note, cities are not stagnant but move in accordance with changes in their leadership and economic structure. Mayors come and go, as do business executives, and conditions change. Cities may be in a transitional phase or they may transition between different types of regimes. But as we see these 13 cities over the past 30 years, the tendency is toward a whittling of city hall's political base (decline of parties, competition from suburbs) and away from business concentration (diversified urban economies, mobile capital). Much as we may interpret and debate the meaning of our 13 city profiles, the direction of change is toward some kind of pluralist or hyperpluralist system.

It is the big city mayor who is left to pick up and pull together the pieces.

The paste of this mayoral personality can hold groups together temporarily despite the divisive pulls of hyperpluralism. Among the centrifugal forces of the system—the fragmented pieces within the city, the external pulls from other governments, and the vacuum created by the decline of parties and local business—the mayor often stands as the only person capable of developing a focus.

When Stone (1988, 1989) discusses urban regimes, he in fact points to these mayor-centered coalitions. Stone's conception of "power to do something" is about the capacity to put people together; mayors can gather scattered splinters with the aim of promoting a limited endeavor.

Our 13 cities provide many examples of successful coalitions that have been more firmly hitched to mayoral personality than to any institution. So

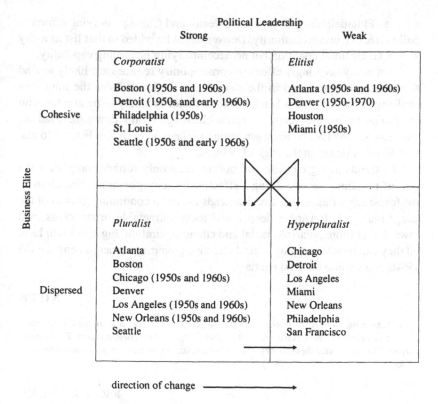

Political Leadership

Figure 15.1 Conceptualizing Political Regimes in Thirteen Cities
Thanks go to Ron Vogel for the idea for this figure. For another example see Ronald Vogel, *Local Political Economy* (1992), Gainesville: University of Florida Press.

it has been with Andrew Young's Atlanta, Coleman Young's Detroit, Art Agnos' San Francisco, Harold Washington's Chicago, Vincent Schoemehl's St. Louis, Tom Bradley's Los Angeles, and Charles Royer's Seattle.

The dependence upon mayors has its down side, too. Once the mayor disappears or loses public confidence, so too may the coalition. Harold Washington's death precipitated an immediate collapse of his coalition; Wilson Goode's loss of credibility in Philadelphia exacerbated that city's fragmentation; Coleman Young's coalition in Detroit probably will last only as long as Young's tenure in office.

With each coalition collapse, too, the task of assembling its successor may grow more difficult. For many of the big cities profiled here, the fragmented political environment of the 1980s has prevented the formation of any kind of effective governing regime. Centrifugal forces now appear to control such cities as Philadelphia, Miami, New Orleans, and Chicago, defying efforts to build effective central authority. Denver might be added to that list as a city within an electoral coalition but no accompanying governing capability.

Even many seemingly effective contemporary regimes are likely to find their authority threatened in the near future. St. Louis faces the imminent challenge of an emerging black electoral majority likely to be at odds with traditional business leaders; a similar challenge looms down the road for Houston; on the Pacific Rim, governing regimes in both San Francisco and Los Angeles teeter on the edge of collapse.

The trends in big city life and politics may only further complicate the task of building—or rebuilding—effective big-city governing coalitions in the foreseeable future. As America heads toward a continuing growth of the exogenous city, toward a deeper and more intractable urban crisis, and toward ever more polarized racial and ethnic separation, big cities will have all they can handle in trying to find effective governing arrangements for the 1990s and on into the 21st century.

NOTE

1. Like most models, these oversimplify the actualities of the case as well as the more complicated conceptions of elitist, pluralist, hyperpluralist, and corporatist theory. For heuristic purposes we have boiled these concepts down to the relative capacities of political and business leaders.

REFERENCES

Banfield, E. (1961). Political influence. New York Free Press.
Banfield, E. (1965). Big city politics. New York: Random House.
Bernstein, R. (1988, February 29). 20 years after the Kerner Report: Three societies, all separate. New York Times, p. B8.
Connery, R., & Leach, R. (1960). The federal government and metropolitan areas. Cambridge, MA: Harvard University Press.
Dahl, R. A. (1961). Who governs? New Haven: Yale University Press.
Federal Bureau of Investigation. (1960). Crime in the United States: Uniform crime reports for the United States. Washington, DC: U.S. Department of Justice.
Federal Bureau of Investigation. (1989). Crime in the United States: Uniform crime reports for the United States. Washington, DC: U.S. Department of Justice.
Gelfand, M. (1975). A nation of cities. New York: Oxford University Press.
Harris, F. R., & Wilkins, R. W. (1988). Quiet riots: Race and poverty in the United States. New York: Pantheon.
Hollingshead, A. B. (1949). Elmtown's youth. New York: John Wiley.
Hunter, F. (1953). Community power structure: A study of decision makers. Garden City, NY: Anchor.
Kantor, P. (1988). The dependent city: The changing political economy of urban America. Glenview, IL: Scott, Foresman.

Logan, J. R., & Molotch, H. L. (1987). *Urban fortunes: The political economy of place*. Berkeley: University of California Press.

Lowi, T. J. (1969). *The end of liberalism: Ideology, policy, and the crisis of public authority*. New York: Norton.

Lynd, R. S., & Lynd, H. M. (1929). *Middletown: A study in American culture*. New York: Harcourt, Brace & World.

Lynd, R. S., & Lynd, H. M. (1937). *Middletown in transition: A study in cultural conflicts*. New York: Harcourt, Brace & World.

Mills, C. W. (1959). *The power elite*. New York: Oxford University Press.

Mollenkopf, J. H. (1983). *The contested city*. Princeton, NJ: Princeton University Press.

National Advisory Commission on Civil Disorders. (1968). *Report*. New York: Dutton.

National Journal Reports, July 28, 1973.

New York Times/CBS News Poll, 1978.

Racial attitudes in fifteen American cities. (1968). University of Michigan, Survey Research Center.

Savitch, H. V. (1979). *Urban policy and the exterior city: Federal, state and corporate impacts upon major cities*. New York: Pergamon.

Savitch, H. V. (1988). *Post-industrial cities: Politics and planning in New York, Paris and London*. Princeton, NJ: Princeton University Press.

Sayre, W. S., & Kaufman, H. (1961). *Governing New York City: Politics in the metropolis*. New York: Norton.

Schmitter, P. (1977). *Comparative political studies*. Beverly Hills, CA: Sage.

Simmie, J. (1981). *Power, property and corporatism*. London: Macmillan.

Sternlieb, G., & Hughes, J. (1975). *Post industrial America*. New Brunswick, NJ: Rutgers University Press.

Stone, C. M. (1988). *Political change and regime continuity in postwar Atlanta*. Paper presented at the Annual Meeting of the American Political Science Association, Washington, DC.

Stone, C. N. (1989). *Regime politics: Governing Atlanta, 1946-1988*. Lawrence, KS: University Press of Kansas.

Truman, D. B. (1960). *The governmental process: Political interests and public opinion*. New York: Knopf.

U.S. Bureau of the Census. (1960). *Compendium of city government finances*. (G-CF60-No. 2). Washington, DC: Government Printing Office.

U.S. Bureau of the Census. (1962). *County and city data book*. Washington, DC: Government Printing Office.

U.S. Bureau of the Census. (1968). *City government finances in 1968*. (GF, No. 4). Washington, DC: Government Printing Office.

U.S. Bureau of the Census. (1976). *City government finances in 1976*. (GF, No. 4). Washington, DC: Government Printing Office.

U.S. Bureau of the Census. (1984). *City government finances in 1984*. (GF, No. 4). Washington, DC: Government Printing Office.

U.S. Bureau of the Census. (1988a). *City government finances in 1988*. (GF, No. 4). Washington, DC: Government Printing Office.

U.S. Bureau of the Census. (1988b). *County and city data book*. Washington, DC: Government Printing Office.

Vidich, A. J., & Bensman, J. (1958). *Small town in mass society*. Princeton, NJ: Princeton University Press.

Yates, D. (1977). *The ungovernable city: The politics of urban problems and policy making*. Cambridge, MA: MIT Press.

Name Index

Subject Index

About the Contributors

CAROLYN TEICH ADAMS has been a member of the faculty at Temple University for 15 years, teaching courses in urban public policy, economic development, housing, and program evaluation. She has authored numerous articles and books, including an award-winning cross-national study, *Comparative Public Policy: The Politics of Social Choice in Europe and America*, now in its third edition. In 1988 she published *The Politics of Capital Investment*, a study of Philadelphia's pattern of investment in public facilities and the impact of those facilities on the city's neighborhoods. Her other work on Philadelphia includes a coauthored book on the city's political and economic history since World War II, published in 1991 in the Temple University Press Comparative Cities Series.

PHILLIP L. CLAY is a distinguished scholar of the American city. An undergraduate honors graduate of the University of North Carolina at Chapel Hill, he holds a Ph.D. in City Planning from MIT, where he now teaches courses on housing, urban demographics, and community development. From 1980 to 1984, he served as Assistant Director of the MIT-Harvard Joint Center of Urban Studies. He has lectured widely and made presentations to professional and community groups. He is author of *Neighborhood Renewal* (1979) and (with Robert Hollister) *Neighborhood Planning and Politics* (1983).

RICHARD E. DeLEON is Professor of Political Science, Urban Studies, and Public Administration at San Francisco State University. His work focuses on urban social movements and growth control politics in U.S. cities. He is writing a book on San Francisco politics.

JOE R. FEAGIN received his Ph.D. from Harvard University and recently became Graduate Research Professor in Sociology at the University of Florida, Gainesville. His research interests include urban sociology, racial and ethnic relations, social problems, and public policy. He is the author or coauthor of more than 90 research articles and monographs and more than 20 books, including *Free Enterprise City: Houston in Political and Economic Perspective* (1988). Feagin's *Ghetto Revolts* was nominated for a Pulitzer Prize. Presently, he is at work on a book with Mel Sikes, featuring hundreds of in-depth interviews with African Americans, demonstrating the continuing significance of race in American society.

BARBARA FERMAN is Associate Professor of Political Science at Illinois Institute of Technology. She was a Visiting Assistant Professor at Barnard College, Columbia University. From 1983 to 1985, she was director of research for the New York City Community Development Agency. She is author of *Governing the Ungovernable City: Political Skill, Leadership and the Modern Mayor* (1985) and coauthor of *The Political Hand: Policy Implementation and Youth Employment Programs* (1985). She is author of numerous articles and papers on urban politics, public policy, and implementation. She is currently writing a book on growth machine politics and political change in Pittsburgh and Chicago. Ferman is a member of the editorial boards of *Urban Affairs Quarterly* and *Policy and Politics*. She holds a Ph.D. in political science from Brandeis University (1983).

ARNOLD FLEISCHMANN is Associate Professor of Political Science at the University of Georgia. He received his Ph.D. from The University of Texas. His research has concentrated on the politics of urban development and has appeared in the *Journal of Politics, Public Administration Review, Social Science Quarterly, Urban Affairs Quarterly*, and other journals.

ANDREW D. GLASSBERG is Associate Professor of Political Science and Public Policy Administration at the University of Missouri-St. Louis. He holds a Ph.D. in Political Science from Yale University. His main research interest is in urban administration. Together with colleagues at the University of Missouri-Columbia and the University of Missouri-Kansas City, he serves as editor of the *American Review of Public Administration*. His research involves both American and comparative urban administration. He has studied local governments and their response to budgetary problems in the United States and United Kingdom. He is the author of *Representation and Urban Community*, a study of neighborhood government in London, and *Public Administration Review* articles on urban fiscal crises in the United States. He is currently studying the utilization of citizen evaluation of public services by the City of St. Louis and administrative responses to privatization in British local government.

MARGARET T. GORDON is Dean and Professor at the Graduate School of Public Affairs at the University of Washington. Before assuming the deanship in 1988, she was Director of the Center for Urban Affairs and Policy Research and Professor of Journalism and Sociology at Northwestern University. She coedited *Urban Affairs Quarterly* from 1981 to 1985. In 1989 she published *The Female Fear* with her colleague Stephanie Riger. The book, which focuses attention on the role of society's institutions in women's fear as it relates to rape, was selected by the American Library Association as one of the most important books of 1989. It will be published in paperback in 1991. Her other research examines the relationship of the news media to public policy. In addition to studying the impacts of news coverage of violent crime, including rape, Gordon and her colleagues have done an extensive study of news investigations and their influence on public policy. Their work has appeared in several scholarly journals and will be published in 1991 as *The Journalism of Outrage*. She received her Ph.D. from Northwestern University.

DENNIS JUDD is Professor and Chair in the Department of Political Science, University of Missouri-St. Louis. He has published extensively on urban political economy, urban revitalization, and public policy. Recent books include *The Politics of American Cities; The Development of American Public Policy* (with David Brian Robertson); *Regenerating the Cities* (coedited with Michael Parkinson and Bernard Foley); *Leadership and Urban Regeneration* (coedited with Michael Parkinson); and *The Democratic Facade* (with Daniel Hellinger). He is coeditor of the *Urban Affairs Quarterly*. He now is working on a book dealing with enclosed spaces and the new forms of segregation.

HUBERT G. LOCKE is Professor, Graduate School of Public Affairs, University of Washington. He served as Dean of the School from 1982-1988. A native of Detroit, he served as Administrative Assistant to the city's Commissioner of Police from 1966-1967, on the staff of the Wayne County Planning Commission, and the research staff of the Metropolitan Fund. From 1972-1976, he was Dean of the College of Public Affairs and Associate Professor of Urban Studies at the University of Nebraska at Omaha. He is the author of *The Detroit Riot of 1967* and of numerous articles and essays on urban policing in America.

LAURIE McCUTCHEON is president of McCutcheon Demographic Research, a firm she began in 1986. The firm's focus is on applying sociodemographic and economic information to issues important to the public and private sectors. McCutcheon conducts research on demographic issues as they relate to such areas as labor force shortages, migration, education, health care, housing, and growth. She is currently completing a study of the causes and consequences of declining homeownership in the Seattle area in order to help design private-sector initiatives to respond to the decline in affordable housing. Previously, she has conducted research on the changing family in Washington State and changes in the labor force in the area. She is affiliated with the Graduate School of Public Affairs, Institute for Public Policy and Management at the University of Washington. She received her Ph.D. in Sociology with a specialty in demography from Brown University in 1977.

ROBERT E. PARKER received his Ph.D. from The University of Texas and is Assistant Professor of Sociology at the University of Nevada, Las Vegas. His research interests include urban sociology, the sociology of work, and race and ethnic relations. Recently, he co-authored *Building American Cities* (1990) with Joe R. Feagin. He has authored several other chapters on Houston dealing with such aspects as political change and economic restructuring. He is currently engaged in an analysis of the social costs stemming from rapid urbanization in southern Nevada, and is working on a manuscript on the temporary help industry and its workers.

WILBUR RICH received his Ph.D. in Political Science from the University of Illinois. Before coming to Wayne State, he held positions with the Illinois Department of Mental Health, Connecticut Department of Mental Health, and Columbia University. He has taught at the University of Illinois, University of Michigan, and University of Wisconsin. He has acted as consultant, researcher, and program evaluator for a variety of public-sector agencies and local governments. His primary areas of research are municipal personnel policy, urban politics, and school politics. He is the author of *The Politics of Urban Personnel Policy: Reformers, Politicians, and Bureaucrats* (1982) and *Coleman Young and Detroit Politics: From Social Activist to Power Broker* (1989). In addition, he has published several articles and reports concerning local government administrative problems. He currently serves on the editorial board of Wayne State University Press and three academic journals. Professor Rich was also one of the winners of Wayne State University's Career Development Chair Award for the academic year 1989-1990.

ALAN L. SALTZSTEIN is Professor of Political Science and Coordinator of Public Administration programs at California State University, Fullerton. He holds a Ph.D. in Political Science from UCLA. He is the author of *Public Employees and Policymaking* and several articles dealing with urban politics and personnel policymaking in cities. His recent research has included an analysis of the effect of Proposition 13 on budgetary decision making, and he has authored several articles with Raphael Sonenshein on Los Angeles city politics.

H. V. SAVITCH is Professor of Urban Policy at the College of Urban and Public Affairs, University of Louisville. He has authored books and written scholarly articles in urban affairs. His books include *Urban Policy and the Exterior City* (Pergamon, 1979) and *Post Industrial Cities* (Princeton, 1988). He has directed federally sponsored projects on the planning and administration of urban communities and has been a consultant to the current mayor of New York City, David Dinkins. This former Fulbright Scholar has lectured and been a resident researcher at the University of Paris, the Ecole Normale Superieur, the Institute of Urban and Physical Planning (Warsaw Polytechnic), and the London School of Economics.

RAPHAEL J. SONENSHEIN is an Associate Professor of Political Science at California State University, Fullerton. His writings on interracial politics have appeared in the *Western Political Quarterly, the Political Science Quarterly,* and in several book chapters. He has coauthored several studies about Los Angeles politics and government with Alan Saltzstein and is currently completing a book on biracial coalition politics in Los Angeles.

WILLIAM B. STAFFORD is the Executive Director of the Trade Development Alliance of Greater Seattle. Between 1971 and 1990, he was with the City of Seattle in a number of positions, including Deputy Mayor to Charles Royer and Director of the Office of Intergovernmental Relations. He was responsible for the national, state, regional, and international relations of the city. He is a graduate of Colorado College and has an MBA from Columbia University. He is an Adjunct Assistant Professor at the Graduate School of Public Administration of the University of Washington. He has been involved in a number of national, state, and regional urban advisory committees, chairing the National Urban Consortium among the big cities on urban research in 1980. He was chosen by the Seattle-King County Municipal League as outstanding public official in 1986.

GENIE N. L. STOWERS is Assistant Professor of Political Science and Public Affairs at the University of Alabama at Birmingham. She has done extensive research on ethnic politics and ethnic groups' ability to influence urban politics; much of this work has focused upon Cuban-Americans in Miami. She also works, within an urban context, on the financing and delivering of human services.

JOHN CLAYTON THOMAS is Professor of Public Administration and Director of the L. P. Cookingham Institute of Public Affairs at the University of Missouri Kansas City. His work has appeared in *Public Administration Review, Urban Affairs Quarterly,* the *American Journal of Political Science,* the *Journal of Urban Affairs,* and other journals. He is also the author of *Between Citizen and City: Neighborhood Organizations and Urban Politics in Cincinnati.*

RONALD K. VOGEL is an assistant professor of political science at the University of Louisville. His articles have appeared in *Urban Affairs Quarterly,* the *Journal of Urban Affairs,* and *Economic Development Quarterly.* He is author of a forthcoming book on local political economy. His research focuses on the dynamics of business-government relations in the community and how this affects the quality of life.

ROBERT K. WHELAN is Professor and Associate Dean in the College of Urban and Public Affairs at the University of New Orleans. He is coauthor of *Urban Policy and Politics in a Bureaucratic Age,* and he has authored (or coauthored) numerous articles, book chapters, and papers. His current research interests center on the politics of urban economic development in New Orleans and in Canadian cities.

CARTER WHITSON is a doctoral candidate in the Political Science Department of the University of Missouri-St. Louis. His scholarly interest is in urban political economy and policy. "Denver: Boosterism Versus Growth" is his first published work.

ALMA H. YOUNG is Professor of Urban and Regional Planning and Director of the Ph.D. Program in Urban Studies, College of Urban and Public Affairs at the University of New Orleans. She has been actively involved in planning initiatives and other civic activities in the City of New Orleans. Her research areas include the political economy of urban development, women and development, and Caribbean political developments. She is a former president of the Caribbean Studies Association and the coeditor of *Militarization in the Non-Hispanic Caribbean./ (1986).*

Printed in the United States
By Bookmasters